John M. Emmerich

SHELDON

HANDBOOKS OF AMERICAN NATURAL HISTORY

ALBERT HAZEN WRIGHT, ADVISORY EDITOR

Aquatic Plants of the United States

BY WALTER CONRAD MUENSCHER

AQUATIC PLANTS
of the United States

BY

WALTER CONRAD MUENSCHER

Late Professor of Botany
in the New York State College of Agriculture
at Cornell University

Comstock Publishing Associates
A DIVISION OF
CORNELL UNIVERSITY PRESS
ITHACA AND LONDON

Copyright 1944 by Comstock Publishing Company, Inc.;
Copyright © 1972 by Mrs. W. C. Muenscher

All rights reserved. Except for brief quotations in a review, this book, or parts thereof, must not be reproduced in any form without permission in writing from the publisher. For information address Cornell University Press, 124 Roberts Place, Ithaca, New York 14850.

First published 1944
Sixth printing 1976

Published in the United Kingdom by Cornell University Press Ltd., 2-4 Brook Street, London W1Y 1AA.
International Standard Book Number 0-8014-0306-5
Printed in the United States of America

TO

MY WIFE

MINNIE WORTHEN MUENSCHER

PREFACE

MONOGRAPHS have been prepared for only a few groups of aquatic plants of the United States. Treatments of all the aquatic plants of any considerable area are few. Floras and catalogs frequently reveal a paucity of information about the aquatic phanerogams and their distribution. The experiences which have led to the preparation of this book began thirty years ago when an attempt was made to classify the aquatic plants of San Juan Island in the state of Washington. Then, and on many occasions since, it was found that in a study of the vegetation of a region the vascular aquatic plants frequently received the least consideration.

Since 1914 the writer has had opportunities to observe, study, and collect aquatic plants in every state of the United States of America. During this time the continent has been crossed eighteen times. Many states have been visited several times, and many of the larger lakes and streams have been examined for their vegetation. The principal types of vegetation in the tidal waters and coastal marshes and ponds have also been studied. More than ten thousand sheets of herbarium specimens have been collected.

The present book is largely based upon these field studies of vascular aquatic plants and the specimens collected. It includes most of the vascular plants growing in the waters of the United States. Submersed and emersed species of fresh, brackish, and salt waters are treated.

During the last fifteen years the writer has been fortunate in receiving assistance in the making of field studies and the collecting of specimens from several students of aquatic plants. The enthusiasm and interest of these men have made it a pleasure to wade many a swamp or morass and to navigate more than one thousand lakes and ponds in order to dredge or trawl from their depths specimens revealing characteristics that are seldom found in the herbarium. The following men were especially helpful and their services are greatly appreciated: Dr. P. R. Burkholder, Dr. W. E. Manning, Dr. A. A. Lindsey, Dr. Bassett Maguire, Dr. R. T. Clausen, Dr. O. F. Curtis, Jr., Dr. O. L. Justice, Mr. W. T. Winne, Dr. Duane Isely, and Mr. R. A. Thorne.

The illustrations are nearly all from original drawings made directly from the plants. They were made under the direction of the writer by Miss Elfriede Abbe and Mr. Nickolas Nickou. The sources of the twelve illustrations which have been used with permission are indicated under their legends. The following figures have been made available by the authors indicated: the original, unpublished plate of *Najas* seeds for figure 28 is kindly loaned by R. T. Clausen; figure 25, A–F is from W. A. Setchell; figure 50, C–F is from Rydberg; figure 87, B–E is from Parija; figure 101 is from an unpublished thesis by Miss Hilda Elena Aboy; figures 122–123 are from W. T. Winne. Permission to reproduce these is gratefully acknowledged. The following six figures are taken from previous publications by the author: figures 7, 20, and 21, G–F are from the *Annals of Botany;* figure 102 is from the *American Journal of Botany;* figures 32 and 133 are from *Poisonous Plants of the United States,* published by The Macmillan Company. Acknowledgment is due the publishers for permission to use these figures here.

The author expresses his sincere appreciation and thanks to his associates at Cornell University for many kindnesses and willing help: to Dr. R. T. Clausen for comments, advice, and criticism of certain groups; to Dr. Duane Isely for reading the manuscript and testing most of the keys in the laboratory and field; and to Miss Babette I. Brown for her assistance with many tedious details of the preparation of the final manuscript, the reading of the proof, and the compilation of the index.

WALTER C. MUENSCHER

Ithaca, New York
June, 1944

CONTENTS

	PAGE
INTRODUCTION	1
Scope of this book	1
Distribution of aquatic plants	3
Reproduction of aquatic plants	4
Vegetative propagation	4
Fruit and seed production	6
Storage and treatment of seeds for planting	7
Weight of seeds	9
KEY TO THE FAMILIES	11
AQUATIC PLANTS ARRANGED BY FAMILIES	15
Typhaceae (Cattail Family)	15
Sparganiaceae (Bur-Reed Family)	18
Potamogetonaceae (Pondweed Family)	27
Najadaceae (Naiad Family)	65
Juncaginaceae (Arrow-Grass Family)	72
Alismaceae (Water-Plantain Family)	78
Butomaceae (Flowering-Rush Family)	95
Hydrocharitaceae (Frogbit Family)	101
Gramineae (Grass Family)	112
Cyperaceae (Sedge Family)	140
Araceae (Arum Family)	175
Lemnaceae (Duckweed Family)	181
Mayacaceae (Bog-Moss Family)	189
Eriocaulaceae (Pipewort Family)	191
Xyridaceae (Yellow-eyed-Grass Family)	196
Pontederiaceae (Pickerelweed Family)	199

CONTENTS

	PAGE
Juncaceae (Rush Family)	206
Marantaceae (Arrowroot Family)	213
Polygonaceae (Buckwheat Family)	213
Chenopodiaceae (Goosefoot Family)	218
Amaranthaceae (Amaranth Family)	223
Caryophyllaceae (Pink Family)	226
Ceratophyllaceae (Hornwort Family)	227
Nymphaeaceae (Water-Lily Family)	231
Ranunculaceae (Buttercup Family)	246
Cruciferae (Mustard Family)	252
Podostemaceae (Riverweed Family)	255
Crassulaceae (Stonecrop Family)	257
Rosaceae (Rose Family)	258
Callitrichaceae (Water-Starwort Family)	258
Elatinaceae (Waterwort Family)	262
Lythraceae (Loosestrife Family)	265
Onagraceae (Evening-Primrose Family)	269
Trapaceae (Water-Nut Family)	274
Haloragidaceae (Water-Milfoil Family)	277
Umbelliferae (Parsley Family)	287
Primulaceae (Primrose Family)	298
Menyanthaceae (Buck-Bean Family)	301
Hydrophyllaceae (Waterleaf Family)	306
Scrophulariaceae (Figwort Family)	307
Lentibulariaceae (Bladderwort Family)	320
Acanthaceae (Acanthus Family)	329
Plantaginaceae (Plantain Family)	329
Lobeliaceae (Lobelia Family)	332
Compositae (Composite Family)	334
Isoëtaceae (Quillwort Family)	339
Equisetaceae (Horsetail Family)	345
Marsileaceae (Water-Clover Family)	346
Salviniaceae (Water-Fern Family)	350
Parkeriaceae (Floating-Fern Family)	352
GLOSSARY	355
INDEX	365

INTRODUCTION

Scope of this book

The concept of aquatic plants is subject to various interpretations. It is difficult to make a definition that can be rigidly adhered to in actual practice. There are too many borderline species that may or may not be included. Aquatic plants, as interpreted in this book, are those species which normally start in water and must grow for at least a part of their life cycle in water, either completely submersed or emersed. A few borderline species of bogs and marshes are also included.

Almost everyone can recall a swamp, a field, or even a desert region which has been converted into a lake. The terrestrial plants of such an area sometimes persist and grow, but only for a short time; they never become aquatics. Anyone may observe on the shore of a lake, pond, or stream certain terrestrial plants that are periodically inundated or immersed; they grow in the water for a brief period but they are not aquatics. Before such plants become adapted to an aquatic habitat or succumb, the water usually recedes, and they return to their terrestrial habits. Such plants are not included in this book.

Many aquatic plants are variable in their response to fluctuations in the water level which may range from complete submersion at floodwater to stranding during receding of the water or in severe drought. Many species, therefore, even though they start in water, must be plastic enough to undergo a period of existence as land forms in addition to their existence as submersed plants and sometimes also as intermediate emersed plants. All of these states or forms are here treated under the name of the species, usually a binomial. There appears to be no justification for using five or six additional Latin tri- or quadrinomials in discussing the variations of such variable species as *Potamogeton gramineus* or *Sagittaria latifolia*. To honor these usually unstable, environmental forms with taxonomic recognition is of doubtful value to the average person interested in aquatic plants.

Woody plants have been omitted from this book. None are true aquatics. Some may start in the water but they also thrive if stranded. Many may grow in areas that are inundated for longer or shorter periods every year. The more common and striking species occurring under such conditions are: *Cephalanthus occidentalis* L., Buttonbush; *Taxodium ascendens* Brongn., Pond cypress; *T. distichum* Rich., Bald cypress; *Rhizophora mangle* L., Mangrove; *Nyssa aquatica* Marsh., Tupelo gum; *Planera aquatica* Gmel., Water elm; *Myrica gale* L., Sweet gale; *Chamaedaphne calyculata* Moench., Leatherleaf; and *Salix* spp., Willows.

The general plan of this book includes a key to the families with aquatic species, followed by a more detailed treatment of the several families. For each family there is included a brief description, a key to the genera treated whenever more than one is discussed, a description of each genus, a key to the aquatic species in each genus whenever more than one is listed, and a statement of the general habitat and range of each species. The descriptions of the families and genera usually conclude with a statement indicating the number of genera and species recognized for the world. Most of the species are illustrated. A map is provided for nearly every species to indicate its known distribution by states. These maps have been prepared from field observations, specimens in herbaria, and trustworthy published data. Generalized statements of range in manuals compiled a generation or so ago have been avoided. The monographs of the following authors treating the special groups indicated were of help in the preparation of the distribution maps of the species: A. S. Hitchcock, grasses; R. T. Clausen, *Najas;* A. Beetle, *Scirpus;* H. K. Svenson, *Eleocharis;* K. Mackenzie, *Carex;* J. G. Smith, *Sagittaria;* M. L. Fernald, *Potamogeton;* E. C. Ogden, *Potamogeton;* L. Benson, *Ranunculus;* F. W. Pennell, *Scrophulariaceae;* N. E. Pfeiffer, *Isoëtes.*

The aquatic plants of the United States are nearly all indigenous species; the following, however, represent introduced species that have become naturalized within at least some waters. The source of each species is indicated after its name: *Potamogeton crispus,* Europe; *Najas minor,* Europe; *Sagittaria montevidensis,* South America; *Butomus umbellatus,* Europe; *Hydrocleis nymphoides,* the tropics; *Limnocharis flava,* South America; *Anacharis densa,* South America; *Manisuris altissima,* the tropics; *Pholiurus incurvus,* Europe; *Eichhornia cras-*

sipes, the tropics; *Eichhornia paniculata,* Brazil; *Nasturtium officinale,* Europe; *Lythrum salicaria,* Eurasia; *Trapa natans,* Asia; *Myriophyllum brasiliense,* South America; *Nymphoides peltatum,* Europe; *Veronica beccabunga,* Europe; *Cotula coronopifolia,* Africa; *Marsilea quadrifolia,* Europe. Several of these species were introduced intentionally as ornamental aquarium or garden-pool plants. Others were brought in fortuitously.

Distribution of aquatic plants

The distribution of aquatic plants is frequently considered to be rather cosmopolitan. Compared to a mesophytic habitat the hydrophytic habitat is often less subject to fluctuations in temperature and water supply; but the dissolved salts and nutrients and the color and transparency of the water, as well as the physical and chemical properties of the water and also the bottom, are subject to much variation even in waters with but slight differences in altitude or latitude. Some species of aquatic plants tolerate a wide range of variation in habitat. These have a cosmopolitan distribution as is illustrated by the twenty species tabulated in Table 1. Every one of these has been observed in each of six large lakes located in various regions of the northern United States. All of these species are sufficiently mobile to become widely disseminated. If observations and conclusions regarding distribution are limited to such tolerant species, it is easy to see how the impression may be gained that aquatic plants have a cosmopolitan distribution.

Certain species of aquatic plants, however, are rather exacting in their requirements and consequently may be somewhat restricted in their range. Their limited distribution may be due to the temperature or depth of the water, the physical properties of the bottom, the reaction of the water or bottom, the quantity or quality of the salts dissolved in the water, the competition of other plants, and other factors. In Table 2 the known distributions of twenty-five restricted species are tabulated for eleven lakes in New York state. The lakes, selected from a relatively limited area, show extreme ranges in altitude, reaction of the water, type of bottom, and degree of salinity. Most of the species are known from only a few, or from a single one, of the lakes selected. It is evident from the examples cited in Tables 1 and 2 that, depending upon their tolerance and aggressiveness and also upon their mobility, aquatic plants, like terrestrial plants, contain many restricted species as well as many cosmopolitan ones.

TABLE 1

The distribution of 20 cosmopolitan species of aquatic plants in 6 widely separated lakes in the northern United States

(x = Known from the lake)

Species	Lake					
	Champlain (Vt.–N.Y.)	Cayuga (N.Y.)	Erie (Ohio)	Flathead (Mont.)	Pend d'Oreille (Idaho)	Whatcom (Wash.)
Typha latifolia	x	x	x	x	x	x
Potamogeton richardsonii	x	x	x	x	x	x
" gramineus	x	x	x	x	x	x
" natans	x	x	x	x	x	x
" epihydrus	x	x	x	x	x	x
" pectinatus	x	x	x	x	x	x
" pusillus	x	x	x	x	x	x
" zosteriformis	x	x	x	x	x	x
Najas flexilis	x	x	x	x	x	x
Alisma plantago-aquatica	x	x	x	x	x	x
Sagittaria latifolia	x	x	x	x	x	x
Scirpus acutus	x	x	x	x	x	x
" americanus	x	x	x	x	x	x
Lemna minor	x	x	x	x	x	x
" trisulca	x	x	x	x	x	x
Spirodela polyrhiza	x	x	x	x	x	x
Ceratophyllum demersum	x	x	x	x	x	x
Myriophyllum exalbescens	x	x	x	x	x	x
Utricularia vulgaris	x	x	x	x	x	x
Bidens beckii	x	x	x	x	x	x

Reproduction of aquatic plants

In their natural environment aquatic angiosperms reproduce and spread by seeds and in many species also by vegetative propagation. Many of them blossom and fruit in abundance in shallow water but seldom mature seeds when growing in deep water or where they are continuously submersed.

Vegetative propagation

In numerous species vegetative propagation is accomplished by special organs; in others any part of the stem may break off and take root. The most common propagating organs are rhizomes, runners,

TABLE 2

The distribution of 25 aquatic plants of restricted range in 11 selected lakes in New York State

(x = Known from the lake)

Species	Fort Pond	Orange Lake	Highland Lake	Lake Tear	Saranac Lake	Long Lake	Tupper Lake	Cayuga Lake	Cayuta Lake	Chautauqua Lake	Lake Ontario
Sparganium fluctuans					x	x	x				
" minimum				x	x						x
Potamogeton capillaceus	x										
" diversifolius						x					
" filiformis								x		x	x
" praelongus								x	x	x	x
" spirillus	x	x	x		x	x	x		x	x	x
" vaginatus								x			
" vaseyi									x	x	x
Najas gracillima					x						
" olivacea								x			
Alisma gramineum											x
Anacharis occidentalis	x	x						x			x
Juncus militaris	x		x								
Cabomba caroliniana		x									
Subularia aquatica						x	x				
Elatine minima	x		x								
Hippuris vulgaris								x			
Myriophyllum tenellum	x				x	x	x			x	x
Nymphoides cordatum	x					x	x				
Utricularia purpurea	x					x	x				
" resupinata						x					
Justicia americana								x			x
Littorella americana						x				x	
Lobelia dortmanna			x		x	x	x				

tubers, corms, bulbils, stems, leaves, and shortened axillary or terminal leafy axes, the "winter buds." Examples of plants illustrating these types of vegetative propagation are listed below.

Rhizomes. *Acorus, Distichlis, Eleocharis, Nelumbo, Nuphar, Phragmites, Potamogeton, Scirpus, Sparganium, Spartina, Thalassia, Typha,* and *Zostera.*

Runners. *Alternanthera, Berula, Calla, Decodon, Echinodorus,*

Hydrocleis, Justicia, Nymphoides, Pistia, Polygonum, Potamogeton, Ranunculus, and *Veronica.*

Tubers. *Eleocharis, Potamogeton, Scirpus, Sagittaria,* and *Vallisneria.*

Corms. *Alisma* and *Damasonium.*

Vegetative buds in leaf axils. *Butomus, Juncus,* and *Lysimachia.*

Stems. *Anacharis, Cabomba, Najas, Nasturtium,* and *Myriophyllum.*

Leaves. *Nymphoides* and *Rorippa.*

Winter buds. *Brasenia, Ceratophyllum, Potamogeton,* and *Utricularia.*

Fruit and seed production

The seeds of aquatic plants are borne in various types of fruits. The morphological units or entities which separate from the parent plant and contain an embryo plant, however, are not always just seeds. Although commonly called "seeds" many of them are actually fruits, that is, they are seeds surrounded by a pericarp. In the classification which follows, the genera treated in this book are grouped according to the morphological nature of their "seeds." Group 1 contains genera with true seeds produced in capsules which at maturity usually dehisce. Group 2 includes genera with berry-like fruits, with one, few, or many seeds, in which the pericarp, though indehiscent, finally may decompose sufficiently to allow the seeds to become free. Group 3 contains genera with indehiscent fruits, usually 1-seeded, generally referred to as "seeds." The common types include the achene, grain, utricle, nutlet, mericarp, and nut.

1. Fruit a capsule or capsule-like structure, at maturity usually dehiscing to discharge the several or many naked seeds.
   Bramia, Butomus, Decodon, Eichhornia, Elatine, Eriocaulon, Glaux, Gratiola, Halophila, Hemianthus, Herpestis, Heteranthera, Hottonia, Howellia, Hydrocleis, Hydrolea, Hydrotrida, Juncus, Jussiaea, Justicia, Limnobium, Limnocharis, Limosella, Lindernia, Lobelia, Ludvigia, Lysimachia, Lythrum, Macuillamia, Mayaca, Menyanthes, Mimulus, Nasturtium, Nymphoides, Plantago, Podostemum, Rorippa, Scheuchzeria, Spergularia, Subularia, Thalassia, Tillaea, Utricularia, Veronica, and *Xyris.*
2. Fruit a berry or berry-like organ, indehiscent, but seeds finally free.
   a. Many-seeded — *Nuphar, Nymphaea,* and *Vallisneria.*
   b. Few-seeded—*Acorus, Anacharis, Cabomba, Calla, Peplis,* and *Pistia.*
   c. One-seeded — *Orontium, Peltandra,* and *Thalia.*

3. Fruit indehiscent, 1-seeded (1- to 3-seeded in *Sparganium* and *Brasenia*); mature seed surrounded by the pericarp.
   a. Achene — *Alisma, Bidens, Carex, Ceratophyllum, Cotula, Cymodocea, Cyperus, Damasonium, Dulichium, Echinodorus, Eleocharis, Eriophorum, Halodule, Jaumea, Lilaea, Littorella, Lophotocarpus, Najas, Phyllospadix, Polygonum, Pontederia, Potamogeton, Potentilla, Ranunculus, Rynchospora, Ruppia, Sagittaria, Scirpus, Sparganium, Typha,* and *Zannichellia.*
   b. Grain — *Alopecurus, Beckmannia, Calamagrostis, Catabrosa, Distichlis, Echinochloa, Fluminea, Glyceria, Hydrochloa, Leersia, Luziola, Manisuris, Monanthochloë, Panicum, Paspalum, Phalaris, Pholiurus, Phragmites, Pleuropogon, Reimarochloa, Spartina,* and *Zizania.*
   c. Utricle — *Acnida, Alternanthera, Lemna, Salicornia, Spirodela, Suaeda, Wolffia, Wolffiella,* and *Zostera.*
   d. Nutlet — *Callitriche, Hippuris, Myriophyllum, Proserpinaca,* and *Triglochin.*
   e. Mericarp — *Berula, Eryngium, Hydrocotyle, Lilaeopsis, Oenanthe, Oxypolis,* and *Sium.*
   f. Nut — *Nelumbo* and *Trapa.*

Storage and treatment of seeds for planting

Although autumn is the time when most aquatic plants naturally disperse their seeds, attempts to establish plants by seeds often are more successful if the seed is planted in the spring. The reasons for this are several. The species to be planted are either absent or rare in the locality, in consequence of which seeds planted in autumn are frequently destroyed by waterfowl; whereas in the late spring many of the waterfowl have moved on, and native vegetation supplies more food for other wild life. Seeds planted in autumn may be carried away by currents, winds, or the movement of ice into deeper water; they may be cast upon the shore or buried under bottom sediment. Between autumn and spring the water level often undergoes great changes with the result that the seeds may be left in places unsuited for their growth. It is apparent, therefore, that seeds which can be stored until spring without losing their vitality and are then planted have a greater chance to germinate and develop into mature plants. The seeds of many aquatic plants germinate poorly or erratically if they are dried before planting. Others, such as the seeds of *Zizania aquatica, Orontium aquaticum, Peltandra virginica,* and *Vallisneria americana,* are killed if they are air-dried. In general, seeds of aquatic plants can be kept in a viable condition until planting time or for several years if they are stored in water at or just above freezing temperature, 1–3 degrees C. Table 3 shows some results of germination tests made with seeds of several aquatic plants after they had been stored in water for one, three, or five years.

TABLE 3

Germination of seeds of aquatic plants stored in water at 1–3° C for from 1 to 5 years

Species	Per cent of germination after		
	1 year	3 years	5 years
Sparganium americanum	16	85	93
S. eurycarpum	12	91	84
S. multipedunculatum	—	92	—
Butomus umbellatus	61	64	68
Potamogeton berchtoldii	71	82	59
Anacharis occidentalis	29	46	—
Vallisneria americana	89	64	23
Alisma plantago-aquatica	92	85	94
Sagittaria latifolia	48	81	92
Zizania aquatica	94	—	—
Nymphaea odorata	51	92	—
Nuphar variegatum	48	87	80
Utricularia geminiscapa	19	38	—
U. purpurea	42	53	—
Nasturtium officinale	93	82	68

REFERENCES

Crocker, William. Germination of seeds of water plants. Bot. Gaz. 44:375–380. 1907.

———, and W. E. Davis. Delayed germination in seed of *Alisma Plantago.* Bot. Gaz. 58:285–321. 1914.

Duvel, J. W. T. The germination and storage of wild rice seed. U. S. Dept. Agr., Bur. Plant Indus. Bull. 90:1–13. 1906.

Edwards, T. I. The germination and growth of *Peltandra virginica* in the absence of oxygen. Torrey Bot. Club Bull. 60:573–581. 1933.

Guppy, H. B. On the postponement of the germination of seeds of aquatic plants. Royal Physical Soc. Edinburgh. Proc. 13:344–360. 1894–1897.

Ludwig, F. Ueber durch austrocknen bedingte Keimfähigkeit der Samen einiger Wasserpflanzen. Biol. Centralbl. 6:299–300. 1886.

Muenscher, W. C. The germination of seeds of *Potamogeton*. Ann. Bot. 50:805–822. 1936.

———. Storage and germination of seeds of aquatic plants. Cornell Univ. Agr. Exp. Sta. Bull. 652. 1936.

Weight of seeds

The "seeds" of aquatic plants vary from minute, in several genera in which numerous seeds are produced in a capsule, to large, in the nut-like seeds of the Lotus and Water nut. The following list records the number of dry "seeds" estimated to be contained in one pound for one hundred species of aquatic plants. The "seeds" of *Trapa natans* are about one million times as heavy as those of *Utricularia cornuta*.

Species	Number of seeds in one pound	Species	Number of seeds in one pound
Sparganiaceae:		*Scheuchzeria palustris*	70,000
Sparganium americanum	50,000	*Triglochin maritima*	600,000
S. angustifolium	250,000	Alismaceae:	
S. chlorocarpum	55,000	*Alisma plantago-aquatica*	700,000
S. eurycarpum	10,000	*Sagittaria cuneata*	900,000
S. fluctuans	120,000	*S. engelmanniana*	600,000
S. minimum	250,000	*S. latifolia*	1,000,000
S. multipedunculatum	80,000	*S. rigida*	700,000
Potamogetonaceae:		Butomaceae:	
Potamogeton confervoides	200,000	*Butomus umbellatus*	2,500,000
P. epihydrus	120,000	Hydrocharitaceae:	
P. filiformis	200,000	*Anacharis occidentalis*	250,000
P. foliosus	400,000	*Vallisneria americana*	900,000
P. gramineus	300,000	Gramineae:	
P. illinoensis	80,000	*Glyceria borealis*	900,000
P. natans	90,000	*G. striata*	2,000,000
P. nodosus	80,000	*Zizania aquatica* (green)	15,000
P. pectinatus	100,000	*Z. aquatica* (dry)	25,000
P. praelongus	30,000	Cyperaceae:	
P. pusillus	420,000	*Cladium mariscoides*	250,000
P. richardsonii	130,000	*Eleocharis calva*	200,000
P. spirillus	600,000	*E. palustris major*	400,000
Ruppia maritima	500,000	*E. quadrangulata*	200,000
Zannichellia palustris	700,000	*Eriophorum virginicum*	1,000,000
Zostera marina	80,000	*Scirpus acutus*	400,000
Najadaceae:		*S. americanus*	163,000
Najas flexilis	700,000	*S. fluviatilis*	80,000
N. gracillima	900,000	*S. paludosus*	175,000
N. marina	150,000	*S. robustus*	150,000
N. muenscheri	800,000	*S. smithii*	500,000
Juncaginaceae:		*S. validus*	500,000
Lilaea subulata	800,000		

Species	Number of seeds in one pound	Species	Number of seeds in one pound
ARACEAE:		CALLITRICHACEAE:	
Acorus calamus	70,000	Callitriche hermaphroditica	4,000,000
Calla palustris	150,000	C. heterophylla	4,000,000
Orontium aquaticum	2,500	C. stagnalis	2,000,000
ERIOCAULACEAE:		LYTHRACEAE:	
Eriocaulon septangulare	3,500,000	Decodon verticillatus	400,000
XYRIDACEAE:		Lythrum salicaria	10,000,000
Xyris caroliniana	50,000,000	ONAGRACEAE:	
X. smalliana	35,000,000	Ludvigia alternifolia	15,000,000
PONTEDERIACEAE:		L. palustris	130,000
Heteranthera reniformis	14,000,000	TRAPACEAE:	
Pontederia cordata	20,000	Trapa natans (dry)	200
JUNCACEAE:		T. natans (green)	75
Juncus acuminatus	30,000,000	HALORAGIDACEAE:	
J. militaris	20,000,000	Hippuris vulgaris	500,000
J. pelocarpus	35,000,000	Myriophyllum heterophyllum	250,000
POLYGONACEAE:		M. humile	3,000,000
Polygonum amphibium	200,000	Proserpinaca palustris	80,000
P. coccineum	130,000	UMBELLIFERAE:	
P. hydropiperoides	260,000	Lilaeopsis lineata	300,000
CHENOPODIACEAE:		PRIMULACEAE:	
Suaeda maritima	400,000	Lysimachia terrestris	900,000
AMARANTHACEAE:		MENYANTHACEAE:	
Acnida cannabina	200,000	Menyanthes trifoliata	200,000
CERATOPHYLLACEAE:		Nymphoides peltatum	250,000
Ceratophyllum echinatum	30,000	SCROPHULARIACEAE:	
NYMPHAEACEAE:		Gratiola lutea	40,000,000
Brasenia schreberi	40,000	Hemianthus micranthemoides	25,000,000
Nuphar advena	18,000	Limosella subulata	9,000,000
N. polysepalum	30,000	Lindernia dubia	50,000,000
N. variegatum	35,000	LENTIBULARIACEAE:	
RANUNCULACEAE:		Utricularia cornuta	175,000,000
Ranunculus aquatilis	1,500,000	U. resupinata	135,000,000
R. flabellaris	350,000	U. vulgaris	12,000,000
CRUCIFERAE:		LOBELIACEAE:	
Nasturtium officinale	2,500,000	Lobelia dortmanna	10,000,000
ROSACEAE:			
Potentilla palustris	1,200,000		

KEY TO THE FAMILIES

1. Plants without roots, floating or submersed.
   2. Stems slender, leafy, floating, or the lower part anchored on the bottom.
      3. Leaves in whorls, without bladders .... CERATOPHYLLACEAE
      3. Leaves alternate or rarely whorled, usually some of them bearing bladders .... LENTIBULARIACEAE
   2. Stems not developed; plant reduced to a small, undifferentiated, flat, globose or tubular, floating frond .... LEMNACEAE
1. Plants with roots.
   4. Plants free floating.
      5. Plants reduced to 1 or a few small, flat, floating fronds .... LEMNACEAE
      5. Plants with several to many leaves inserted on an axis.
         6. Leaves simple.
            7. Axis horizontal; leaves not over 2 cm. long, not in a rosette ....
            .... SALVINIACEAE
            7. Axis vertical; leaves 4 to 10 cm. long, erect, in a rosette ARACEAE (*Pistia*)
         6. Leaves pinnately-compound or lobed, more than 10 cm. long, attached to a vertical axis .... PARKERIACEAE
   4. Plants rooted on the bottom, submersed or emersed.
      8. Reproduction by spores borne in sporangia.
         9. Leaves compound or at least pinnately-lobed, alternate.
            10. Leaves palmately-compound, with 4 entire leaflets ..... MARSILEACEAE
            10. Leaves pinnately-compound, pinnatifid or lobed ...... PARKERIACEAE
         9. Leaves simple, often reduced.
            11. Leaves whorled, reduced to minute scales and fused into a sheath; stems jointed, hollow; sporangia in a terminal cone .... EQUISETACEAE
            11. Leaves quill-like, crowded into a basal tuft; sporangia on the inner side of the leaf base .... ISOËTACEAE
      8. Reproduction by flowers and seeds.
         12. Leaves parallel-veined, simple (except in a few forms with fleshy spadix [Araceae]); stem bundles mostly scattered when viewed in cross-section .... (MONOCOTYLEDONS)
            13. Leaves in whorls at the numerous nodes of the stem ....
            .... HYDROCHARITACEAE (*Anacharis*)

13. Leaves not in whorls.
  14. Ovary superior.
    15. Pistil solitary, simple, or compound.
      16. Ovary 1-celled or appearing so.
        17. Fruit 6- to many-seeded.
          18. Leaves in a basal tuft .... XYRIDACEAE
          18. Leaves alternate along the stem.
            19. Leaves more than 3 cm. long .. PONTEDERIACEAE (*Heteranthera*)
            19. Leaves less than 2 cm. long .... MAYACACEAE
        17. Fruit 1- or 3-seeded.
          20. Flowers solitary in the axils of leaves .... NAJADACEAE
          20. Flowers in clusters.
            21. Flowers crowded on a fleshy spadix .... ARACEAE
            21. Flowers not on a fleshy spadix.
              22. Plants acaulescent annuals, with basal terete leaves; flowers in a scapose spike and some axillary in the basal leaves .... JUNCAGINACEAE (*Lilaea*)
              22. Plants perennial.
                23. Plants submersed, marine or rarely in brackish water .... POTAMOGETONACEAE
                23. Plants emersed, mostly of freshwater or marshes.
                  24. Perianth showy; leaves cordate .... PONTEDERIACEAE (*Pontederia*)
                  24. Perianth wanting or inconspicuous; leaves linear.
                    25. Leaves in 3 ranks; stem solid, with closed leaf sheaths .... CYPERACEAE
                    25. Leaves in 2 ranks.
                      26. Inflorescence composed of spikelets; stems mostly hollow, jointed .... GRAMINEAE
                      26. Inflorescence a spike or head.
                        27. Flowers in a thick, solitary terminal spike .... TYPHACEAE
                        27. Flowers in burlike heads .... SPARGANIACEAE
      16. Ovary 2- or 3-celled.
        28. Flowers in a fleshy spadix; leaves often broad or netted-veined .... ARACEAE
        28. Flowers not in a fleshy spadix.
          29. Flowers with showy perianth; leaves broad, with inflated petiole .... PONTEDERIACEAE (*Eichhornia*)
          29. Flowers not showy; leaves grasslike; petioles wanting or not inflated.
            30. Inflorescence a head or spike terminating a naked scape; capsule 2- or 3-seeded .... ERIOCAULACEAE
            30. Inflorescence an open or crowded cymose cluster on a leafy stem; capsule many-seeded .... JUNCACEAE
    15. Pistils several or many, free.
      31. Perianth wanting or inconspicuous.

32. Pistils 3 to 6; fruit 1- or 2-seeded . . . . JUNCAGINACEAE
   32. Pistils 4, if fewer then raised on a long stipe in fruit; fruit 1-seeded
        . . . . POTAMOGETONACEAE
  31. Perianth showy.
   33. Pistils 1-ovuled, forming achenes . . . . ALISMACEAE
   33. Pistils with many ovules, forming many-seeded follicles BUTOMACEAE
 14. Ovary inferior.
   34. Flowers irregular; fruit 1-seeded . . . . MARANTACEAE
   34. Flowers regular; fruit several- to many-seeded . . . . HYDROCHARITACEAE
 12. Leaves netted-veined, simple, or compound; stem bundles in a ring when viewed in cross-section . . . . (DICOTYLEDONS)
  35. Flowers with several or many separate pistils.
   36. Leaves opposite.
    37. Blade simple . . . . CRASSULACEAE
    37. Blade finely dissected . . . . NYMPHAEACEAE (*Cabomba*)
   36. Leaves alternate.
    38. Leaves peltate . . . . NYMPHAEACEAE
    38. Leaves not peltate.
     39. Leaves pinnately compound . . . . ROSACEAE
     39. Leaves simple or palmately compound, divided, or lobed . . . .
        . . . . RANUNCULACEAE
  35. Flowers with a solitary pistil.
   40. Pistil 1-ovuled; fruit 1-seeded.
    41. Fruit a large spiny nut . . . . TRAPACEAE
    41. Fruit an achene or utricle.
     42. Flowers in involucrate heads; corolla tubular . . . . COMPOSITAE
     42. Flowers not in involucrate heads; corolla wanting or if present not tubular.
      43. Leaves with stipular sheaths . . . . POLYGONACEAE
      43. Leaves without stipules or sheath.
       44. Leaves opposite or whorled.
        45. Leaves whorled . . . . HALORAGIDACEAE (*Hippuris*)
        45. Leaves opposite.
         46. Leaves reduced to fleshy scales . . . .
             . . . . CHENOPODIACEAE (*Salicornia*)
         46. Leaves broad, not scalelike or fleshy . . . .
             . . . . AMARANTHACEAE (*Alternanthera*)
       44. Leaves alternate.
        47. Leaves in a basal cluster; flowers on scapes or basal . . . .
            . . . . PLANTAGINACEAE (*Littorella*)
        47. Leaves scattered along the stem; flowers in panicles or axillary.
         48. Leaves succulent, linear . . . . CHENOPODIACEAE (*Suaeda*)
         48. Leaves not succulent, broad . . . . AMARANTHACEAE (*Acnida*)
   40. Pistil with 2 or more ovules; fruit a capsule or separating into 2 to 4 nutlets.
    49. Plants attached to rocks by disklike processes, cartilaginous, appearing like algae . . . . PODOSTEMACEAE
    49. Plants not attached to rocks.

50. Petals separate or wanting.
  51. Fruit with 6 or more radial cells, large, fleshy .... NYMPHAEACEAE
  51. Fruit with fewer than 6 cells.
    52. Ovary superior; fruit a capsule.
      53. Placenta free-central.
        54. Leaves with stipules; capsule 3-valved .... CARYOPHYLLACEAE
        54. Leaves without stipules; capsule 5-valved .... PRIMULACEAE
      53. Placenta not free-central.
        55. Leaves alternate or basal; placenta parietal .... CRUCIFERAE
        55. Leaves opposite or whorled; placenta axile or basal.
          56. Plants tall, emersed .... LYTHRACEAE
          56. Plants dwarfed, mostly creeping, submersed.
            57. Leaves not linear, glandular-dotted .... ELATINACEAE
            57. Leaves linear, not glandular-dotted ...... LYTHRACEAE (*Peplis*)
    52. Ovary inferior, or if superior then fruit separating into 2 to 4 nutlets.
      58. Fruit a capsule with several seeds .... ONAGRACEAE
      58. Fruit 2- to 4-celled, each cell with 1 seed.
        59. Fruit splitting into 2 one-seeded parts .... UMBELLIFERAE
        59. Fruit separating into 3 or 4 nutlets.
          60. Leaves opposite, simple, entire .... CALLITRICHACEAE
          60. Leaves whorled or alternate, mostly compound, dissected, not entire .... HALORAGIDACEAE
50. Petals united; corolla tubular, bell-shaped, or rotate.
  61. Corolla regular.
    62. Leaves alternate.
      63. Leaves simple.
        64. Blade cordate or peltate .... MENYANTHACEAE (*Nymphoides*)
        64. Blade not cordate or peltate.
          65. Leaves in a basal rosette .... PLANTAGINACEAE (*Plantago*)
          65. Leaves along the stem .... HYDROPHYLLACEAE
      63. Leaves trifoliate or finely dissected.
        66. Leaves trifoliate .... MENYANTHACEAE (*Menyanthes*)
        66. Leaves finely dissected .... PRIMULACEAE (*Hottonia*)
    62. Leaves opposite or whorled.
      67. Leaves whorled; capsule 1-celled .... PRIMULACEAE (*Lysimachia*)
      67. Leaves opposite; capsule 2-celled .... SCROPHULARIACEAE
  61. Corolla irregular, often 2-lipped.
    68. Leaves often with bladders .... LENTIBULARIACEAE
    68. Leaves without bladders.
      69. Leaves alternate or basal.
        70. Plants with milky juice; calyx fused to the ovary .... LOBELIACEAE
        70. Plants without milky juice; calyx free from the ovary ....
        .... SCROPHULARIACEAE
      69. Leaves opposite.
        71. Capsule 2- to 4-seeded .... ACANTHACEAE
        71. Capsule many-seeded .... SCROPHULARIACEAE

AQUATIC PLANTS ARRANGED BY FAMILIES

TYPHACEAE: Cattail Family

Tall, erect, rank, perennial herbs with simple, jointless stems, 2-ranked, linear, sheathing leaves, and thick, branching rootstocks. Inflorescence a dense, rigid spike, 2 to 4 dm. long, usually breaking out of a spathe. Flowers numerous, imperfect, reduced, and naked, the lower pistillate and the upper staminate. Fruit minute, on a long stalk, with a single seed with endosperm. — A single genus with about twelve species, mostly of marshes; several with cosmopolitan distribution.

TYPHA: Cattail

Staminate flowers reduced to a pair of stamens intermixed with hairs; pistillate flowers of single, stipitate pistils with a 1-celled ovary and persistent style with 1-sided stigma, subtended by club-shaped bristles becoming dry at maturity.

NOTE. The staminate portion of the spike soon shrivels after the pollen is shed; the pistillate part of the spike enlarges and becomes 2 to 3 cm. in diameter and up to 4 dm. long. These spikes or "bobs" persist into late winter. They are sometimes gathered for winter decorations. The hairy down is sometimes used for packing material. The leaves of *Typha* are harvested extensively in autumn, stacked, and dried. The ability of the leaves to resist decay, absorb water, and swell makes them well suited for tightening seams to prevent leaks. The butt ends are frequently used for calking barrels, boats, and barges. The upper parts of the leaves are also used for caning or weaving the so-called "flag bottoms" or seats of chairs. The poorer grades of leaves are used for packing nursery stock for shipment.

The rootstocks of *Typha* are rich in starch. They are sometimes eaten

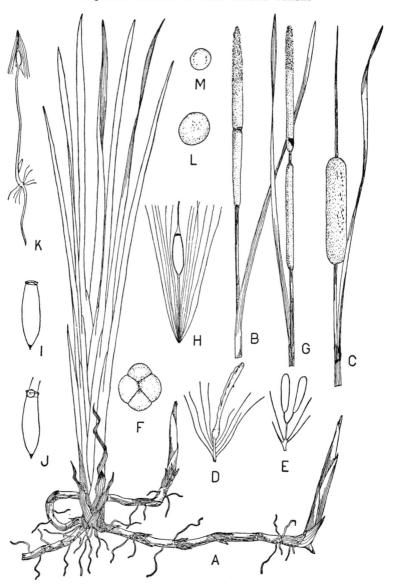

FIG. 1. *Typha latifolia* (A–F), *T. angustifolia* (G–L), and *T. truxillensis* (M).

(A) Young plant with rootstocks; x 1/6. (B) Spike in flower; x 1/5. (C) Spike in fruit; x 1/5. (D) Pistillate flower, x 7. (E) Staminate flower; x 7. (F) Pollen grains, group of four; approximately x 240.

(G) Spike in flower; x 1/5. (H) Mature pistil with bristles; x 5. (I) Fruit; x 10. (J) Fruit with seedling pushing open the lid; x 10. (K) Fruit inverted with protruding seedling; x 3. (L) Pollen grain, single; x 240.

(M) Pollen grain, single; x 240.

by muskrats and other animals and in emergencies have been used for food by man.

KEY TO SPECIES OF TYPHA

1. Staminate and pistillate portions of the spikes contiguous; pistillate flowers without bractlets, with flattened stigma; pollen grains in 4's; lower leaves 10 to 24 mm. broad .... *T. latifolia*
1. Staminate and pistillate portions of the spikes usually distant (not contiguous); pistillate flowers with bractlets and threadlike stigma; pollen grains single; lower leaves 3 to 20 mm. broad.
    2. Bracts subtending the staminate flowers simple or, rarely, forked at apex; hairs among pistillate flowers capitate-thickened .... *T. angustifolia*
    2. Bracts subtending the staminate flowers branched into 3 or more parts at apex; hairs among pistillate flowers capitate .... *T. truxillensis*

Typha angustifolia L. Narrow-leaved cattail. Fig. 1, G–L. Map 1.

Common in bays and marshes both in fresh water and brackish water. Chiefly on lowlands in the eastern and central states.

Typha latifolia L. Common cattail. Fig. 1, A–F. Map 2.

Widespread in shallow bays, sloughs, and marshy and springy places. Throughout the United States.

MAP 1. *Typha angustifolia.*

MAP 2. *Typha latifolia.*

MAP 3. *Typha truxillensis.*

Typha truxillensis HBK. Fig. 1, M. Map 3.

Similar to *T. angustifolia* of which it may represent a large variety;

usually larger and with bluish-green foliage. In brackish marshes along the Atlantic Coast; also reported from Texas and California.

REFERENCE

Fernald, M. L. Rhodora 37: 385. 1935. 42:407. 1940.

SPARGANIACEAE: Bur-Reed Family

Perennial aquatic or marsh plants with alternate, sessile, linear, erect, emersed or floating, 2-ranked leaves, creeping rootstocks, and clustered, fibrous roots. Plants monoecious; the imperfect flowers in globular, sessile or stalked heads on the upper part of the stem or its branches. The upper heads staminate and the lower pistillate. — A single genus with about 15 species, the majority of which are aquatics.

SPARGANIUM: Bur Reed

Pistillate heads few, of many pistils each with a perianth of 3 to 6 sepals. Staminate heads several to many, of numerous flowers of 3 to 5 stamens and minute sepals. Fruits achenes, borne in burlike clusters, at maturity separating, spindle-shaped or cuneate-obovoid, water-disseminated, 1- to 4-seeded. Seeds with a straight embryo and large endosperm (Fig. 1a).

KEY TO SPECIES OF SPARGANIUM

1. Stigmas mostly 2; achenes sessile, obovoid to obpyramidal; sepals nearly as long as the achene; plants erect, about one meter tall; leaves keeled ..... *S. eurycarpum*
1. Stigmas solitary; achenes mostly stalked, tapering; sepals much shorter than the achene; plants smaller, leaves erect or floating.
   2. Stipes and beaks less than 1 mm. long or none; fruiting heads about 1 cm. in diameter; staminate heads mostly solitary .... *S. minimum*
   2. Stipes and beaks each 2 mm. long or more; fruiting heads 1.5 cm. in diameter or larger; staminate heads several.
      3. Beaks strongly curved; stigma short-oblong; achenes reddish-brown; leaves and fruiting stems floating .... *S. fluctuans*
      3. Beaks straight or but slightly curved; stigma linear; achenes green or brownish.
         4. Pistillate heads or branches all axillary; achenes dull; leaves without scarious margins.
            5. Inflorescence simple or the branches strict and bearing 0 to 2 staminate heads .... *S. americanum*
            5. Inflorescence branched; the branches jointed, bearing 3 to 7 staminate heads .... *S. androcladum*
         4. Pistillate heads or branches, at least some, supra-axillary.
            6. Leaves 3 to 4 mm. wide, not scarious-margined, not keeled but usually rounded on back; stem and leaves floating .... *S. angustifolium*

THE PLANTS ARRANGED BY FAMILIES

6. Leaves 3 to 9 mm. wide, with scarious margin near the base, mostly strongly keeled or flat on back; stem and leaves usually erect and emersed.
   7. Beak as long as body of achene; staminate heads 4 to 9 .... *S. chlorocarpum*
   7. Beak shorter than body of achene; staminate heads mostly 1 to 4 .... *S. multipedunculatum*

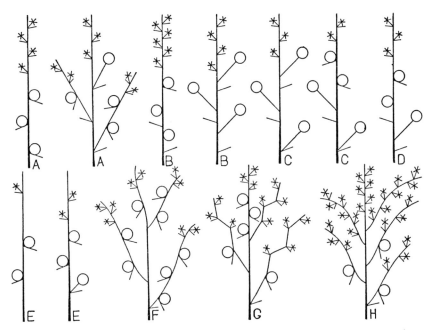

FIG. 1a. *Sparganium*. Diagrams showing relative position of pistillate and staminate heads in the inflorescences of the several species.

O = pistillate head. / = bract. * = staminate head.

(A) *S. americanum*. Pistillate heads axillary or on axillary branches.

(B) *S. chlorocarpum*. Pistillate heads sessile or stalked, at least some of them supra-axillary; staminate heads 4 to 9.

(C) *S. multipedunculatum*. Pistillate heads stalked or the upper sessile, at least some of them supra-axillary; staminate heads 1 to 4.

(D) *S. angustifolium*. Pistillate heads mostly supra-axillary, the lower stalked, the upper sessile.

(E) *S. minimum*. Pistillate heads 2 or 3, axillary, sessile, or the lower short-stalked; staminate heads 1 or 2.

(F) *S. fluctuans*. Branches with axillary pistillate heads below and staminate heads above.

(G) *S. androcladum*. Lower branches zigzag, with 3 to 7 staminate heads and sometimes a solitary lower pistillate head; main axis with sessile, axillary, pistillate heads below and staminate heads above.

(H) *S. eurycarpum*. Lower branches with one pistillate head and several staminate heads; main axis and upper branches with 6 to 10 staminate heads.

Sparganium americanum Nutt. Figs. 2, A–B; 6, E–F. Map 4.

Widespread in shallow ponds, sluggish streams, and sloughs. Throughout the eastern states.

Sparganium andracladum (Engelm.) Morong. Fig. 2, C–D. Map 6.

Along swampy borders of shallow ponds and streams. Chiefly in the northeastern states.

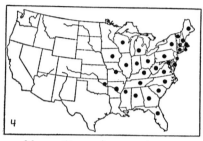
Map 4. *Sparganium americanum.*

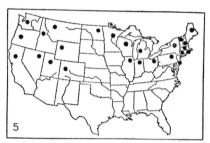
Map 5. *Sparganium angustifolium.*

Sparganium angustifolium Michx. Fig. 3, C–D. Map 5.

In deeper water of ponds, lakes, and streams. Widespread, chiefly in the northern states.

Sparganium chlorocarpum Rydb. Fig. 4, A–B. Map 8.

Common in temporary ponds and on boggy shores. From the northeastern states to the north central states. Extremely short-stemmed specimens of this species have been described as *S. acuale* (Beeby) Rydb. The length of stem varies with the depth of the water; specimens stranded early may be almost stemless.

Map 6. *Sparganium androcladum.*

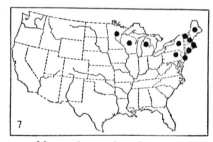
Map 7. *Sparganium fluctuans.*

Sparganium eurycarpum Engelm. Figs. 5, A–C; 6, A–C. Map 9.

In marshes, sloughs, and shallow water. Locally common across the United States except in the South. *S. californicum* Greene and *S. greenei* Morong, based upon specimens from the Pacific Coast with slenderer and more rounded achenes, hardly seem specifically distinct from *S. eurycarpum*.

Sparganium fluctuans (Morong) Robinson. Figs. 3, A–B; 48a, E. Map 7.

In deeper water, mostly in acid ponds and lakes. Chiefly in New England and about the Great Lakes.

Sparganium minimum Fries. Fig. 5, D–E. Map 10.

In cold, spring-fed ponds and sluggish streams, usually on soft bottom. Chiefly in the northeastern states and at higher altitudes from the Rocky Mountains westward.

MAP 8. *Sparganium chlorocarpum.*

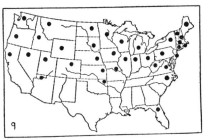
MAP 9. *Sparganium eurycarpum.*

Sparganium multipedunculatum (Morong) Rydb. Figs. 4, C–D; 6, D. Map 11.

Mostly on soft bottom of shallow ponds and stream banks and in sloughs. This is the common species from the northern Rocky Mountain region to the Pacific Northwest; also local in New England.

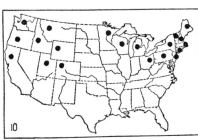
MAP 10. *Sparganium minimum.*

MAP 11. *Sparganium multipedunculatum.*

REFERENCES

Fernald, M. L. Notes on *Sparganium*. Rhodora 24:26–34. 1922.

———. *Sparganium multipedunculatum* in eastern America. Rhodora 27:190–193. 1925.

Fig. 2. *Sparganium americanum* (A, B) and *S. androcladum* (C, D).
(A) Plant; x 1/2. (B) Achene; x 3.
(C) Inflorescence of staminate and pistillate heads; x 1/2. (D) Achene; x 3.

Fig. 3. *Sparganium fluctuans* (A, B) and *S. angustifolium* (C, D).
(A) Plant; x 1/2. (B) Achene; x 3.
(C) Plant and fruiting top; x 1/2. (D) Achene; x 3.

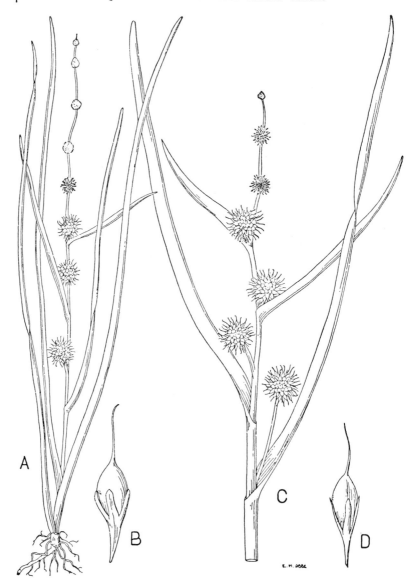

FIG. 4. *Sparganium chlorocarpum* (A, B) and *S. multipedunculatum* (C, D).
(A) Plant; x 1/2. (B) Achene; x 3.
(C) Upper part of plant with staminate and pistillate heads; x 1/2.
(D) Achene; x 3.

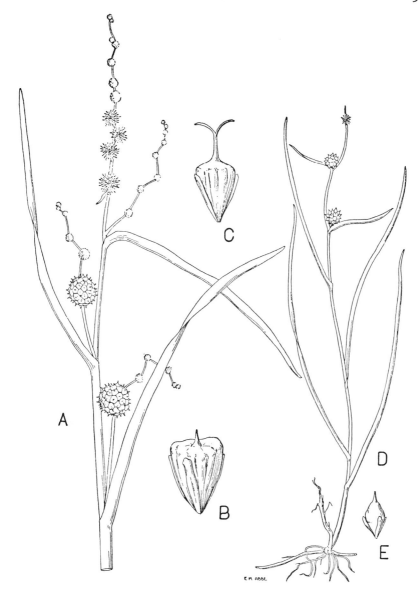

Fig. 5. *Sparganium eurycarpum* (A–C) and *S. minimum* (D, E).

(A) Upper part of plant with staminate and pistillate heads; x 1/4. (B) Achene; x 2. (C) Pistil with 2 style branches; x 2.

(D) Plant; x 1/2. (E) Achene; x 3.

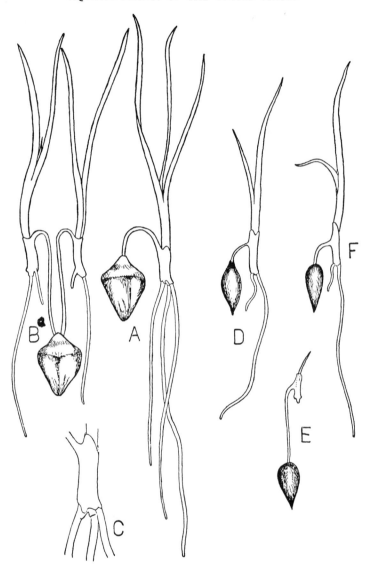

Fig. 6. *Sparganium eurycarpum* (A–C), *S. multipedunculatum* (D), and *S. americanum* (E, F).

(A) Seedling with pericarp still attached; x 3. (B) Pericarp producing two seedlings; x 3. (C) Base of sheath of seedling; x 6.
(D) Seedling attached to pericarp; x 3.
(E, F) Seedling in early and later stages; x 3.

Clausen, R. T. A new variety of *Sparganium americanum*. Rhodora 39:188-190. 1937.
Lakela, Olga. *Sparganium glomeratum* in Minnesota. Rhodora 43:83-85. 1941.

POTAMOGETONACEAE: Pondweed Family

Fresh- or salt-water, submersed aquatic plants, often with creeping rootstocks. Leaves alternate or, rarely, opposite, on erect, jointed stems, or basal, simple, with a sheath or stipulate. Flowers perfect or imperfect, in spikes, cymose clusters, or solitary, often borne in a spathe. Perianth wanting, or of 4 to 6 small, herbaceous or membranous, separate or fused segments; stamens 1 to 6, with extrorse anthers; pistils 1 to 6, distinct or rarely fused; ovary superior; fruit 1-seeded, indehiscent; seeds without endosperm. — This is the largest family of true aquatic seed plants; about 12 genera.

KEY TO GENERA

1. Carpels several, free; plants of fresh or, rarely, of brackish waters.
    2. Flowers perfect; stamens 2 to 4; leaves alternate.
        3. Stamens 4, with petal-like connectives; flowers in spikes; fruit sessile .... *Potamogeton*
        3. Stamens 2, without connectives; flowers not in spikes; fruit on a long stalk .... *Ruppia*
    2. Flowers imperfect; stamen solitary; fruit short-stalked; leaves opposite ....
       ....*Zannichellia*
1. Carpels solitary or fused into a compound pistil; flowers imperfect or, rarely, perfect; plants of marine waters.
    4. Flowers borne in a compressed, 1-sided spike inclosed in a spathe.
        5. Plants dioecious .... *Phyllospadix*
        5. Plants monoecious or, rarely, flowers perfect .... *Zostera*
    4. Flowers borne in the leaf axils.
        6. Leaves flat; stigma solitary .... *Halodule*
        6. Leaves terete or nearly so; stigmas 2 .... *Cymodōcea*

POTAMOGETON: Pondweeds

Aquatics with jointed, leafy stems with fibrous roots from the lower nodes. Leaves 2-ranked, alternate or subopposite, variable, those on the same plant all alike or of two kinds, all submersed or some of them floating. Submersed leaves thin, all linear, reduced to phyllodia, or all broad, from lanceolate to ovate or cordate; floating leaves coriaceous, mostly petioled, with broad, lanceolate to ovate or elliptical blade. Stipules membranous, more or less fused by the margins into a sheath or

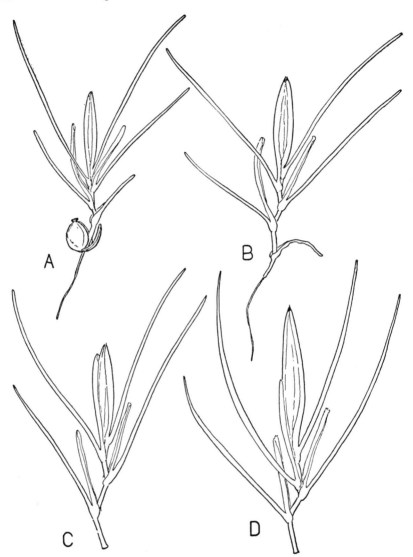

Fig. 7. *Potamogeton berchtoldii.*

(A) Seedling showing primary root 20 days after seed germination. (B) Seedling showing primary root and cotyledon beginning to shrivel, 32 days after germination. (C) Seedling March 2, 1935, 60 days after seed germination; cotyledon and primary root gone, axis resembling a winter bud. (D) Winter bud, as it appeared when taken from a pond on March 2, 1935. All x 2. (From Muenscher, 1936.)

in a few species fused with the base of the blade. Flowers small, crowded in spikes. Spikes sheathed by the stipules in the bud, raised to the surface on long peduncles or submersed on short axillary peduncles. Flowers perfect, regular; perianth of 4 rounded, valvate, greenish lobes, "sepals"; stamens 4, opposite the "sepals"; carpels 4 (rarely 1 to 5), separate, with a short style or the stigma sessile. Fruit drupelike when fresh, more or less compressed, with bony, often keeled endocarp containing a solitary seed with coiled embryo. — The largest genus of true aquatic seed plants of temperate regions; about 60 species.

NOTE. Various interpretations of the flowers of *Potamogeton* have been proposed to explain the nature of the perianth-like parts, the "sepals." In the present description of the inflorescence the term "flower" is used to designate the units of the spike composed of 4 carpels and 4 stamens surrounded by 4 sepaloid structures. This interpretation implies that the sepaloid structures are true sepals fused with the stamens opposite them so that the anthers appear sessile upon them. According to one theory, the sepaloid structure is an expansion of the connective of the stamen. According to another, the flower as here considered is a reduced inflorescence of 4 staminate flowers, each consisting of one stamen and a bract, surrounding a pistillate flower of 4 carpels. According to this view, the spike would be a compound inflorescence. Studies of the vascular anatomy of the flowers supply no evidence to support the theory that the sepaloid structure has arisen from an expanded connective but rather suggests that it represents a perianth segment or bract.

Short branches or abbreviated axes, the so-called "winter buds" or "turions," are produced in the axils of the leaves of some species (Fig. 7, D). These fall to the bottom, take root, and produce new plants. Other species produce long, creeping, branched rootstocks which may terminate in small tubers.

KEY TO SPECIES OF POTAMOGETON

1. Plants with 1 kind of leaves, all submersed.
    2. Leaves broad, lanceolate to elliptical or ovate, never linear, often clasping.
        3. Margin of leaf blades serrulate; winter buds hard, with serrate, rigid, spreading leaves; fruit with long, slender beak ....*P. crispus*
        3. Margin of leaf blades entire, rarely serrulate at tip.
            4. Base of blade tapering, not clasping.
                5. Upper leaves petioled; blades serrulate near apex; plant green ....*P. illinoensis*
                5. Upper leaves sessile or nearly so; blades entire; plant reddish *P. alpinus*
            4. Base of blade clasping.
                6. Blade 10 to 30 cm. long, with cucullate apex; stipules 2 to 8 cm. long, persistent; stem whitish; fruit 4 to 5 mm. long, sharply 3-keeled; embryo with straight apex ....*P. praelongus*
                6. Blade 1 to 12 cm. long, apex not cucullate; stem green; fruit 2 to 4 mm. long, obscurely 3-keeled; embryo with apex curved inward.
                    7. Leaves short, with rounded apex and plain margin, drying dark or olive; stipules small or wanting; peduncle slender ..... *P. perfoliatus*
                    7. Leaves narrowly ovate, with tapering apex and crinkly margin, drying light green; stipules conspicuous, persisting as shreds; peduncle spongy .... *P. richardsonii*
    2. Leaves linear.
        8. Stipules fused with the lower part of the leaf to form a sheath at least 1 cm. long.
            9. Leaves 4 to 8 mm. wide, auricled at base, serrulate, oriented on the axis into a rigid, flattened spray ....*P. robbinsii*
            9. Leaves filiform, rarely up to 3 mm. wide, not auricled, entire, oriented into a lax, diffuse, branched spray.
                10. Stigmas raised on a minute style, capitate; leaves gradually acuminate; rhizomes tuber-bearing ....*P. pectinatus*
                10. Stigmas inconspicuous, broad, and sessile; leaves retuse, blunt, or shortly apiculate.
                    11. Plants short, slender; leaves all filiform; sheaths close around stem; spike with 2 to 5 whorls of flowers ....*P. filiformis*
                    11. Plants coarse, 2 to 5 meters long; leaves on main stem short, flat, their sheaths enlarged to 2 to 5 times the diameter of the stem; spikes with 5 to 12 whorls of flowers ....*P. vaginatus*
        8. Stipules free from the leaf or, rarely, fused to the base for 1 or 2 mm.
            12. Plants with slender, creeping rhizomes; leaves without basal glands.
                13. Peduncles terminal, mostly 5 to 25 cm. long; leaves narrower than the stems, flaccid, filiform, with long, tapering apex .... *P. confervoides*
                13. Peduncles axillary, less than 3 cm. long; leaves broader than the stems, acute or cuspidate at apex ....*P. foliosus*
            12. Plants with short rhizomes or none at all (often rooting at the lower nodes of the stem).

14. Leaves 9- to 35-nerved, subrigid; prominent winter buds with imbricated stipules and ascending blades.
   15. Stems much flattened and winged, about as wide as the leaves; leaves 2 to 5 mm. wide, without basal glands .... *P. zosteriformis*
   15. Stems somewhat flattened, not winged; leaves mostly less than 2 mm. wide, bristle-tipped, with a pair of basal glands ....*P. longiligulatus*
14. Leaves 1- to 7-nerved.
   16. Leaves without basal glands.
     17. Stipules rigid, coarsely fibrous and ciliate, soon disintegrating into fibers ....*P. fibrillosus*
     17. Stipules finely fibrous, not coarsely ciliate ....*P. foliosus*
   16. Leaves, at least some of them, with a pair of basal glands.
     18. Leaves with 5 to 7 nerves, thin; winter buds composed largely of overlapping, whitish, fibrous stipules and blades ....*P. friesii*
     18. Leaves with 3 (rarely 1 or 5) nerves.
        19. Leaves gradually tapering to bristle tips or revolute.
           20. Leaves revolute, rigid; winter buds terminating long slender branches ....*P. strictifolius*
           20. Leaves not revolute; spike 3- to 6-fruited.
              21. Leaves hairlike; winter buds on ascending bristle-like branches; spike cylindric, interrupted, on filiform peduncle ....*P. gemmiparis*
              21. Leaves linear, 3-nerved; spike capitate, on short spreading or recurved peduncle ....*P. hillii*
        19. Leaves obtuse or acute, not bristle-tipped.
           22. Stipules stiffly fibrous or becoming fibrous-fimbriate near the tip.
              23. Leaves suffused with red; peduncle about 1 cm. long; fruit with dorsal keel prominent ....*P. porteri*
              23. Leaves green; peduncle 2 to 6 cm. long; fruit with sharp lateral keels ....*P. clystocarpus*
           22. Stipules membranous or herbaceous, not fibrous.
              24. Body of winter bud 2 to 4 cm. long, covered with scarious stipules ....*P. obtusifolius*
              24. Body of winter bud less than 2 cm. long, solid; leaves green, rarely reddish.
                25. Peduncle filiform, mostly 3 to 8 cm. long; stipules in early stages with edges at least in part connate . *P. pusillus*
                25. Peduncle mostly 0.5 to 3 cm. long; stipules flat or convolute, not connate ....*P. berchtoldii*
1. Plants with 2 kinds of leaves; floating leaves broad and coriaceous; submersed leaves broad and membranous or linear.
26. Submersed leaves broad, never linear.
   27. Floating leaves with 30 to 55 nerves; submersed leaves with 30 to 40 nerves ....*P. amplifolius*

27. Floating leaves with fewer than 30 nerves; submersed leaves with fewer than 30 nerves.
  28. Submersed leaves with more than 7 nerves, all petiolate.
    29. Base of floating leaves cordate or subcordate .... *P. pulcher*
    29. Base of floating leaves tapering or rounded but not cordate .... *P. nodosus*
  28. Submersed leaves mostly with 7 nerves, at least the lower sessile.
    30. Margin of submersed leaves serrulate near apex .... *P. illinoensis*
    30. Margin of submersed leaves entire.
      31. Plant reddish; submersed leaves at least as wide as the floating leaves, mostly on the main stem .... *P. alpinus*
      31. Plant green; submersed leaves narrower than the floating leaves, often numerous on short, axillary branches .... *P. gramineus*
26. Submersed leaves (or phyllodia) linear.
  32. Stipules all free from the leaf bases; spikes of 1 kind only; fruits not at all or but slightly compressed.
    33. Floating leaves more than 1 cm. wide and more than 2 cm. long; winter buds usually wanting.
      34. Submersed leaves tapelike, 2 to 10 mm. wide, with a prominent cellular median band; fruit 3-keeled .... *P. epihydrus*
      34. Submersed leaves terete, often reduced to petiole, mostly less than 1.5 mm. thick, without a median band.
        35. Blade of floating leaves elliptical, with tapering base; fruit 3-keeled, without lateral dimple .... *P. nodosus*
        35. Blade of floating leaves ovate to subcordate; fruit scarcely keeled, with a dimple on each side.
          36. Fruits with concave sides; spikes 3 to 6 cm. long; floating leaves mostly 3 to 10 cm. long .... *P. natans*
          36. Fruits with plane sides; spikes 1 to 3 cm. long; floating leaves 2 to 5 cm. long .... *P. oakesianus*
    33. Floating leaves less than 1 cm. wide and less than 2 cm. long, 5- to 9-nerved.
      37. Submersed leaves filiform, tapering; floating leaves 3 to 8 mm. wide, on marginless petioles; winter buds nearly sessile on short, axillary branches .... *P. vaseyi*
      37. Submersed leaves linear, acute; floating leaves 2 to 4 mm. wide, tapering to margined petioles; winter buds terminating upper branches .... *P. lateralis*
  32. Stipules of all, or at least of some of the lower, leaves fused with the leaf base; winter buds rare; spikes of 2 kinds, those in the axils of the lower submersed leaves globose, submersed on short peduncles; those in the axils of the upper or floating leaves cylindrical, often emersed on longer peduncles; fruit laterally compressed, 3-keeled, with spirally coiled embryo.
    38. Submersed leaves filiform, terminating in a slender thread or bristle tip; floating leaves mostly 3- to 7-nerved, acute or mucronulate.
      39. Fruit with lateral keels low; dorsal keel entire or slightly dentate; fruit with sides nearly flat .... *P. capillaceus*

39. Fruit with lateral keels winged or dentate; dorsal keel coarsely dentate; fruit with a deep dimple on each side .... *P. bicupulatus*
38. Submersed leaves linear, obtuse, or acute, but not tapering into bristle tips; floating leaves mostly with rounded or emarginate apex, 5- to 15-nerved; dorsal keel of fruit usually prominently toothed.
40. Stipules fused for more than ½ their length; submersed leaves blunt; floating leaves slightly oblique and emarginate at apex; fruit with obsolete beak and sides rounded instead of keeled .... *P. spirillus*
40. Stipules fused about ½ their length; submersed leaves pointed; floating leaves rounded, not emarginate at apex; fruit with minute beak and low lateral keels .... *P. diversifolius*

Potamogeton alpinus Balbis (*P. tenuifolius* Raf.). Fig. 9, A–C. Map 12.

Local in cold ponds, lakes, and streams, in shallow or deep water. Northeastern states to the Pacific Northwest.

Potamogeton amplifolius Tuckerm. Broad-leaved pondweed. Figs. 8, A–B; 20, A–E. Map 13.

In deep water of lakes and ponds and in slow streams, sometimes forming extensive beds. Throughout the eastern states except in the extreme South; infrequent on the Pacific Coast.

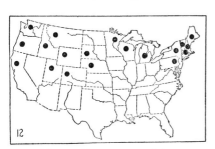
MAP 12. *Potamogeton alpinus.*

MAP 13. *Potamogeton amplifolius.*

Potamogeton berchtoldii Fieber (*P. pusillus* of Fernald). Figs. 10, A–C; 7. Map 14.

Common in shallow ponds, temporary pools, ditches, and sluggish streams. Widespread, especially in the northeastern and central states. This is one of the most ubiquitous aquatics, being among the first to appear in new ponds and in any temporary pool where water collects in a depression. This extremely variable species has many leaf variations that have been designated as subspecies and varieties.

Potamogeton bicupulatus Fernald. Map 15.

Local in ponds and streams. Known from the Allegheny Mountains in Pennsylvania and Tennessee; perhaps of wider distribution.

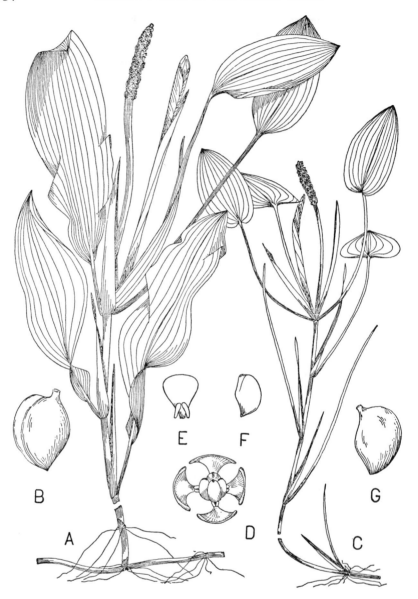

Fig. 8. *Potamogeton amplifolius* (A, B) and *P. natans* (C–G).
(A) Plant; x 1/2. (B) Fruit; x 5.
(C) Plant; x 1/2. (D) Flower; x 4. (E) Connective with 2 anthers; x 5. (F) Carpel; x 5. (G) Fruit; x 5.

THE PLANTS ARRANGED BY FAMILIES 35

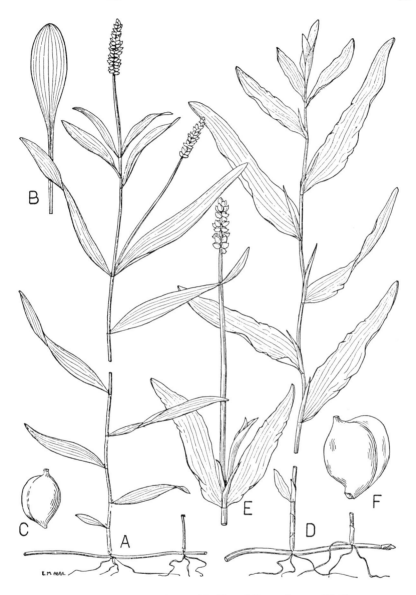

FIG. 9. *Potamogeton alpinus* (A–C) and *P. praelongus* (D–F).
(A) Portion of plant showing habit; x 1/2. (B) A floating leaf; x 1/2. (C) Fruit; x 5.
(D) Portion of small, sterile plant; x 1/2. (E) Upper part of fruiting shoot; x 1/3.
(F) Fruit; x 5.

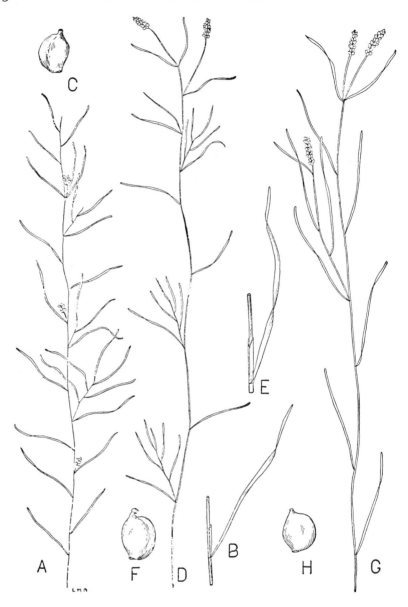

FIG. 10. *Potamogeton berchtoldii* (A–C), *P. pusillus* (D–F), and *P. strictifolius* (G, H).
(A) Part of plant; x 1/2. (B) Leaf with open stipule; x 2. (C) Fruit; x 5.
(D) Part of plant; x 1/2. (E) Leaf with closed stipule; x 2. (F) Fruit; x 5.
(G) Part of plant; x 1/2. (H) Fruit; x 5.

THE PLANTS ARRANGED BY FAMILIES 37

FIG. 11. *Potamogeton capillaceus* (A-C), *P. lateralis* (D, E), *P. diversifolius* (F-H), and *P. gemmiparis* (I, J).

(A) Part of plant; x 1/2. (B) Leaf with stipule; x 2. (C) Fruit; x 5.
(D) Part of plant; x 1/2. (E) Fruit; x 5.
(F) Floating leaf; x 1. (G) Submersed leaf with stipule; x 2. (H) Fruit; x 5.
(I) Part of plant with winter buds; x 1/2. (J) Fruit; x 5.

Map 14. *Potamogeton berchtoldii*.   Map 15. *Potamogeton bicupulatus*.

Potamogeton capillaceus Poiret. Figs. 11, A–C; 21, A–E. Map 16.

Common in ponds and slow streams in regions of sandy or peaty acid soils. Chiefly on the Coastal Plain; from Maine to Florida and Texas.

Potamogeton clystocarpus Fernald. Map 18.

Known only from sluggish pools in the Davis Mountains, Jeff Davis County, Texas.

Potamogeton confervoides Reichenb. Figs. 12, A–B; 20, F–I. Map 18.

Local in acid ponds and slow streams, on sandy or peaty bottom. New England to Pennsylvania and Wisconsin.

Potamogeton crispus L. Fig. 13, A–D. Map 17.

In ponds, small lakes, and sluggish streams; often locally abundant in muddy or polluted streams and bays. Introduced from Europe into eastern North America before 1814. Widespread in the northeastern and north central states; local elsewhere.

Map 16. *Potamogeton capillaceus*.   Map 17. *Potamogeton crispus*.

Potamogeton diversifolius Raf. Fig. 11, F–H. Map 19.

Common in ponds, lakes, and slow streams. Widespread; most abundant in the southern and central states.

Map 18. *Potamogeton confervoides*, ●;
P. *clystocarpus*, ▲.

Map 19. *Potamogeton diversifolius*.

Potamogeton epihydrus Raf. Fig. 14, A–B. Map 20.

Common in streams and spring-fed ponds and near the inlets and outlets of lakes. Most abundant in the northeastern and Great Lakes regions, also local in the Pacific Northwest. Variable with regard to the width of the submersed leaves. Field observations indicate that the leaf width is influenced by the fertility of the substrata in which the plants are growing.

Potamogeton fibrillosus Fernald. Map 21.

In spring-fed pools and small streams, frequently in warm water. Local from the northern Rocky Mountains to eastern Washington and Oregon. Perhaps this may represent only an ecological form of *P. foliosus* Raf.

Map 20. *Potamogeton epihydrus*.

Map 21. *Potamogeton fibrillosus*.

Potamogeton filiformis Pers. Fig. 17, D–E. Map 22.

Local chiefly in shallow water near the shores of calcareous lakes, ponds, and streams. Widespread in the northern and Rocky Mountain states.

Potamogeton foliosus Raf. Fig. 12, G–I. Map 23.

Widely distributed in fresh, calcareous, and brackish waters in lakes, ponds, and slow streams. Throughout the United States. This species

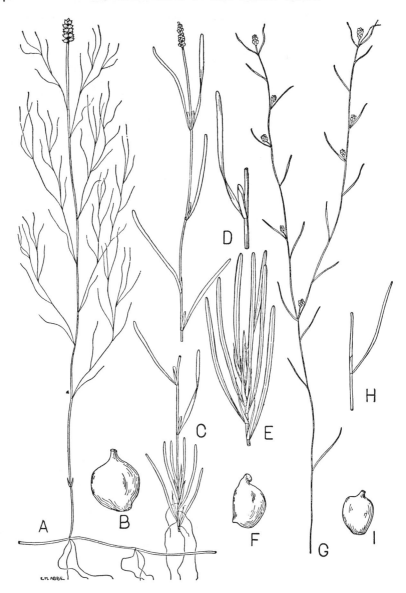

FIG. 12. *Potamogeton confervoides* (A, B), *P. friesii* (C-F), and *P. foliosus* (G-I).
(A) Plant with rootstock; x 1/2. (B) Fruit; x 5.
(C) Part of plant from winter bud; x 1/2. (D) Leaf with stipule; x 1. (E) Winter bud; x 1. (F) Fruit; x 5.
(G) Part of plant; x 1/2. (H) Leaf; x 1. (I) Fruit; x 5.

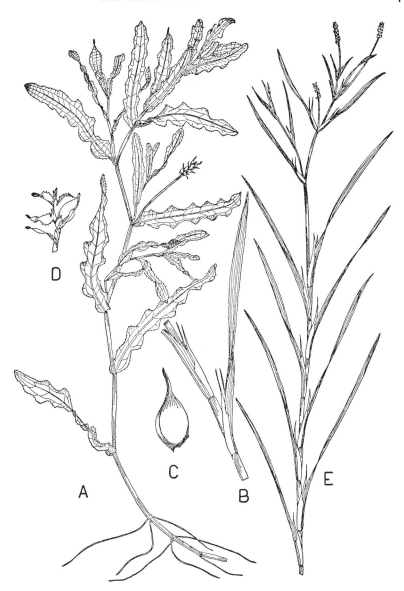

Fig. 13. *Potamogeton crispus* (A-D) and *P. robbinsii* (E).

(A) Plant; x 1/2. (B) Leaf with sheath; x 1. (C) Fruit; x 6. (D) Winter bud or turion; x 1/2.

(E) Portion of plant in flower; x 1/2.

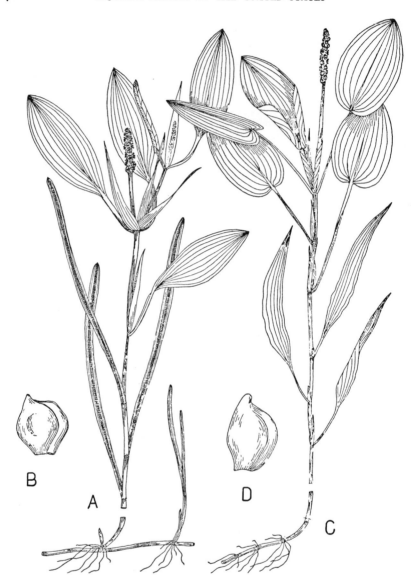

Fig. 14. *Potamogeton epihydrus* (A, B) and *P. pulcher* (C, D).
(A) Plant; x 1/2. (B) Fruit; x 5.
(C) Plant; x 1/2. (D) Fruit; x 5.

THE PLANTS ARRANGED BY FAMILIES 43

FIG. 15. *Potamogeton gramineus* (A, B), *P. vaseyi* (C-E), and *P. spirillus* (F-H).
(A) Upper portion of plant; x 1/2. (B) Fruit, x 5.
(C) Part of plant; x 1/2. (D) Leaf with sheathing stipule; x 2. (E) Fruit; x 5.
(F) Part of plant; x 1/2. (G) Leaf with stipule; x 2. (H) Fruit; x 5.

is represented in the northern states by lower, more branching and bushy plants, *P. foliosus* var. *macellus* Fernald.

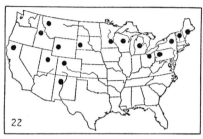

MAP 22. *Potamogeton filiformis*.

MAP 23. *Potamogeton foliosus*.

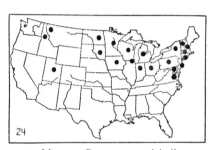

MAP 24. *Potamogeton friesii*.

MAP 25. *Potamogeton gemmiparis*.

Potamogeton friesii Rupr. Fig. 12, C–F. Map 24.

In deep water, mostly in calcareous lakes; also in brackish ponds and slow streams. Chiefly from New York westward to North Dakota.

Potamogeton gemmiparis Robbins. Fig. 11, I–J. Map 25.

Locally common in streams and ponds, chiefly in acid waters. New England.

Potamogeton gramineus L. Figs. 15, A–B; 21, F–J. Map 26.

Common in lakes and slow streams, often in deep water. Widespread across the northern United States. This is an extremely variable species. Its many leaf forms and different habits of branching have been made the basis for a number of varieties and forms.

Potamogeton hillii Morong. Map 27.

Local in ponds and slow streams. New England to Ohio and Michigan.

Potamogeton illinoensis Morong (*P. lucens* of Amer. authors and *P. angustifolius* C. and S.). Fig. 16, C–D. Map 28.

A deepwater plant of lakes and streams, often forming a zone along the edge of currents where streams enter bays of lakes. Most common in the northeastern and north central states; widespread elsewhere.

Map 26. *Potamogeton gramineus.*

Map 27. *Potamogeton hillii.*

Map 28. *Potamogeton illinoensis.*

Potamogeton lateralis Morong. Fig. 11, D–E. Map 29.
Very local in ponds and quiet streams. Massachusetts to Minnesota.

Potamogeton longiligulatus Fernald. Map 30.
Local in calcareous waters of lakes, ponds, and slow streams. Connecticut to Minnesota.

Map 29. *Potamogeton lateralis.*

Map 30. *Potamogeton longiligulatus.*

Potamogeton natans L. Fig. 8, C–G. Map 31.
In shallow lakes, ponds, intermittent pools, and sluggish streams. Widespread, chiefly in the northern states.

Fig. 16. *Potamogeton nodosus* (A, B) and *P. illinoensis* (C, D).
(A) Part of plant and rhizome; x 1/2. (B) Fruit; x 5.
(C) Upper part of plant; x 1/2. (D) Fruit; x 5.

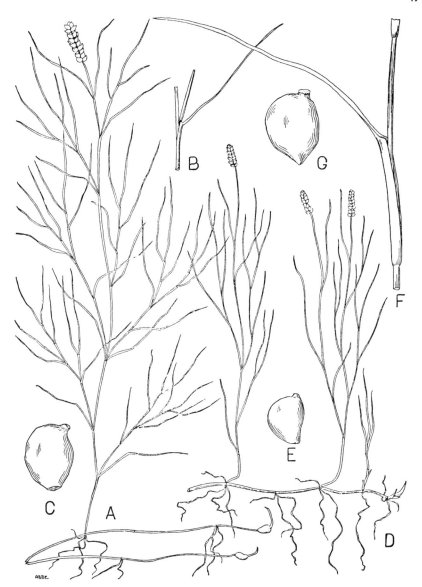

Fig. 17. *Potamogeton pectinatus* (A–C), *P. filiformis* (D, E), and *P. vaginatus* (F, G).

(A) Plant with rootstocks and tubers; x 1/2. (B) Leaf with stipules fused with the sheath; x 1. (C) Fruit; x 5.
(D) Plant showing habit; x 1/2. (E) Fruit; x 5.
(F) Leaf showing enlarged sheath; x 1. (G) Fruit; x 5.

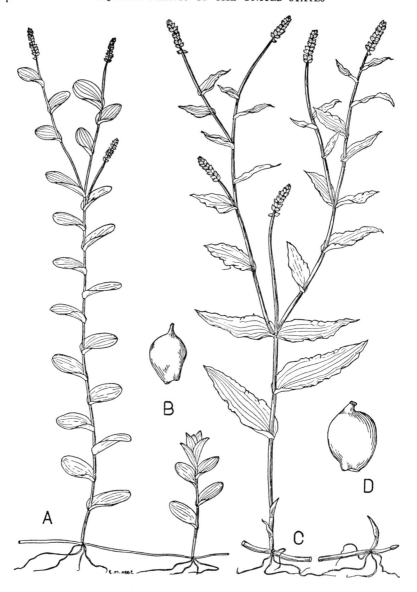

FIG. 18. *Potamogeton perfoliatus* (A, B) and *P. richardsonii* (C, D).
(A) Habit of plant; x 1/2. (B) Fruit; x 5.
(C) Habit of plant; x 1/2. (D) Fruit; x 5.

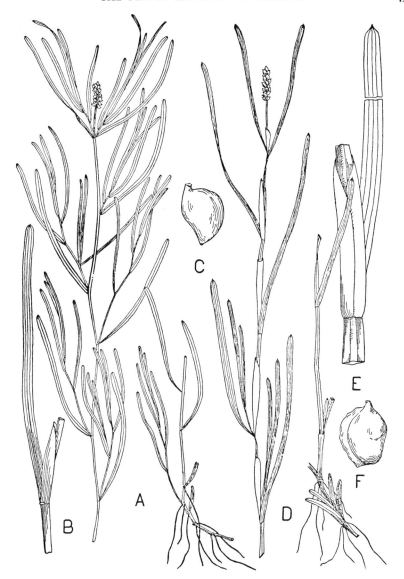

FIG. 19. *Potamogeton obtusifolius* (A–C) and *P. zosteriformis* (D–F).
 (A) Plant; x 1/2. (B) Leaf with sheath; x 2. (C) Fruit; x 5.
 (D) Plant; x 1/2. (E) Leaf with sheath; x 5. (F) Fruit; x 5.

Potamogeton nodosus Poiret (*P. americanus* C. and S.). Fig. 16, A–B. Map 32.

In deep water in rivers and large brooks and near the inlet and outlet channels of lakes. Widespread, nearly throughout the United States.

MAP 31. *Potamogeton natans.*

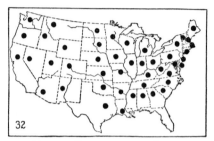

MAP 32. *Potamogeton nodosus.*

MAP 33. *Potamogeton oakesianus.*

MAP 34. *Potamogeton obtusifolius.*

Potamogeton oakesianus Robbins. Map 33.

Mostly in shallow, acid ponds. New England, New York, and Michigan. In the vegetative condition this species has much the general appearance of depauperate specimens of *P. natans*.

Potamogeton obtusifolius Mert. and Koch. Fig. 19, A–C. Map 34.

In shallow ponds and spring-fed streams. Chiefly in the northeastern and Great Lakes regions; infrequent elsewhere.

Potamogeton pectinatus L. Sago pondweed. Fig. 17, A–C. Map 35.

Common in ponds, lakes, and slow streams; chiefly in nonacid waters. Widespread throughout the United States. This species produces edible tubers, fleshy rootstocks, and an abundance of fruits all of which are relished by waterfowl. The tubers and seeds have been planted in many localities to improve feeding places for ducks. These introductions undoubtedly have extended the distribution of this species beyond its natural range.

MAP 35. *Potamogeton pectinatus*.

MAP 36. *Potamogeton perfoliatus*.

Potamogeton perfoliatus L. (*P. bupleuroides* Fernald). Fig. 18, A–B. Map 36.

In brackish waters in ponds and streams; mostly on sandy bottom. Chiefly on the Coastal Plain, from Maine to Florida; local inland.

Potamogeton porteri Fernald. Map 38.

Known only from cold streams in swamps of Lancaster County, Pennsylvania.

Potamogeton praelongus Wulfen. Fig. 9, D–F. Map 37.

In lakes and ponds, usually in deep, clear water but sometimes also in shallow ponds with soft, peaty bottom. Widespread across the northern United States.

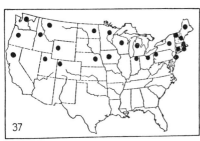

MAP 37. *Potamogeton praelongus*.

MAP 38. *Potamogeton porteri*.

Potamogeton pulcher Tuckerm. Fig. 14, C–D. Map 39.

Shallow ponds, lakes, intermittent pools, and sluggish streams. Chiefly on the Coastal Plain and in the lower Mississippi Valley.

Potamogeton pusillus L. (*P. panormitanus* of Fernald). Fig. 10, D–F. Map 40.

Common in lakes, ponds, and slow streams, often in deep water. Widely distributed throughout the United States.

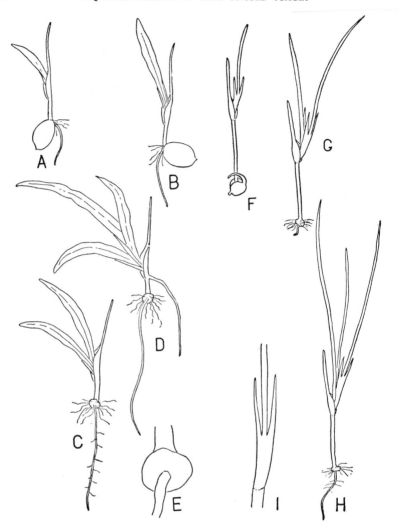

Fig. 20. *Potamogeton amplifolius* (A–E) and *P. confervoides* (F–I).

(A, B, C) Seedlings 3, 8, and 16 days after germination; x 2. (D) Seedling showing both primary and secondary roots, 32 days after germination; x 2. (E) Base of hypocotyl showing "collar"; x 12.

(F, G, H) Seedlings 4, 8, and 15 days after germination; x 2. (I) Base of leaf with stipules; x 4. (From Muenscher, 1936.)

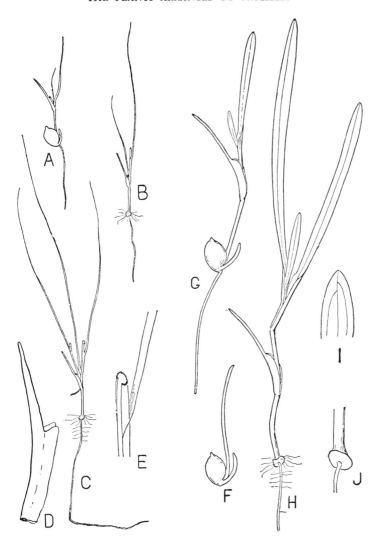

Fig. 21. *Potamogeton capillaceus* (A-E) and *P. gramineus* (F-J).

(A, B, C) Seedling 4, 10, and 28 days after germination; x 2. (D) Cotyledon; x 10. (E) Base of first leaf showing stipular sheath; x 10.

(F, G, H) Seedling 4, 26, and 45 days after germination; x 4. (I) Apex of first leaf; x 12. (J) Base of hypocotyl with "collar"; x 4. (G-J from Muenscher, 1936.)

Note. This species, as now restricted, is usually treated as two varieties, var. *major* including the broad-leaved forms and var. *minor* including the narrow-leaved forms.

There has been much confusion with regard to the small, linear-leaved plants of *Potamogeton* included under the binomial *P. pusillus*. As treated in Gray's *Manual*, *P. pusillus* included a complex group of little understood forms. This complex was segregated into two species by Fernald who retained *P. pusillus* in a restricted sense for those forms with free stipules which are flat but not connate. The rest of the complex, in which the stipules are connate or fused by the edges to form cylinders, were referred by Fernald to *P. panormitanus* Biv. and Bernh. Dandy has pointed out that the plants assigned to *P. panormitanus* Biv. and Bernh. by Fernald should be retained under *P. pusillus* L., and the plants retained under *P. pusillus* by Fernald belong to *P. berchtoldii* Fieber. It is unfortunate that the binomial *P. pusillus* in recent years should have been used first for the whole complex and then only for the segregate with free stipules and finally only for the segregate with connate stipules.

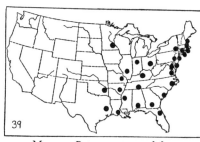 

Map 39. *Potamogeton pulcher*. Map 40. *Potamogeton pusillus*.

Potamogeton richardsonii (Ar. Benn.) Rydb. Fig. 18, C–D. Map 41.

Mostly in deep water of lakes and streams. Widespread, chiefly in the northern United States.

Potamogeton robbinsii Oakes. Fig. 13, E. Map 42.

Abundant in lakes and ponds, often in deep water where it forms the dominant vegetation over large areas. Northeastern and Great Lakes regions, also in the Pacific Northwest.

Potamogeton spirillus Tuckerm. Fig. 15, F–H. Map 43.

In ponds, lakes, and slow streams, mostly on sandy or gravelly bottom. Chiefly in the northeastern and Great Lakes regions.

NOTE. In clear lakes this species may grow completely submersed to a depth of 5 meters, forming low, much-branched plants without floating leaves and with many short, axillary flower spikes which may set fruits in abundance.

MAP 41. *Potamogeton richardsonii.*

MAP 42. *Potamogeton robbinsii.*

Potamogeton strictifolius Ar. Benn. Fig. 10, G–H. Map 44.
Chiefly in calcareous lakes and ponds. Vermont to the Great Lakes and Nebraska.

MAP 43. *Potamogeton spirillus.*

MAP 44. *Potamogeton strictifolius.*

MAP 45. *Potamogeton vaginatus.*

MAP 46. *Potamogeton vaseyi.*

Potamogeton vaginatus Turcz. Fig. 17, F–G. Map 45.
Local in deep water of lakes and rivers. Widespread in the northern states.

Potamogeton vaseyi Robbins. Fig. 15, C–E. Map 46.

In quiet lakes and ponds. From New England to Minnesota and southward to Illinois and Pennsylvania.

Potamogeton zosteriformis Fernald. Figs. 19, D–F; 48a, D. Map 47.

In lakes, ponds, and streams, frequently in deep water; mostly intermixed with larger species of *Potamogeton*. Widely distributed in the northern states except on the Great Plains.

Map 47. *Potamogeton zosteriformis.*

REFERENCES

Morong, Thomas. The Najadaceae of North America. Torrey Bot. Club Mem. 3:1–65. 64 pl. 1893.

Hagström, J. O. Critical researches on the Potamogetons. Kungl. Svenska Vetenskapsakad. Handl. 55: no. 5. 1916.

Fernald, M. L. The linear-leaved North American species of *Potamogeton*, section Axillares. Mem. Amer. Acad. Arts and Sci. 17 (1):1–183. 1932.

Ogden, E. C. The broad-leaved species of *Potamogeton* of North America north of Mexico. Rhodora 45:57–105, 109–163, 171–214. 1943.

Dandy, J. E., and G. Taylor. The typification of *Potamogeton pusillus*. Jour. Bot. (British) 76:90–91. 1938.

Moore, Emmeline. The Potamogetons in relation to pond culture. U.S. Bur. Fisheries Bull. 33:251–291. 1915.

Muenscher, W. C. The germination of seeds of *Potamogeton*. Ann. Bot. 50:805–822. 1936.

Whitford, N. B. Some morphological and anatomical investigations on the flowers of the genus *Potamogeton*. Cornell Univ. Thesis. 1943.

CYMODOCEA

Submersed, dioecious perennials with creeping, jointed rootstocks and basal, linear, terete leaves with acute apex and auricled sheath. Flowers solitary or in cymose clusters; staminate flowers of 2 anthers attached at the same level on a slender stalk; pistillate flowers of 2 fused carpels. Pistil with a slender style terminating in 2 stigmas. Fruit 1-seeded. — About 6 species widespread in tropical marine waters.

Cymodocea manatorum Aschers. Manatee grass. Map 48.

In shallow, salt-water bays and brackish streams. Florida and Texas.

Map 48. *Cymodocea manatorum.*

Map 49. *Halodule wrightii.*

HALODULE

Submersed dioecious perennials with slender, branching, creeping rootstocks bearing fibrous roots at nodes; roots often terminating in fleshy, starchy, tuber-like bodies. Leaves mostly on short, erect, lateral branches, linear, flattened, about 1 mm. wide, the apex unsymmetrical, with a central and two unequal lateral points. Base of leaf with ligulate sheath. Flowers naked; the staminate of 2 anthers attached at different heights near the end of a long stalk; pistillate, a single carpel with short style and solitary stigma. Fruit small, globose. — Two species in tropical marine waters.

Halodule wrightii Aschers. Fig. 22. Map 49.

In shallow waters mostly offshore; vegetative plants are frequently cast on the beach during storms. South Carolina, Florida to Texas.

PHYLLOSPADIX

Submersed dioecious perennials with thick, creeping, and branching rhizomes and branched, slender stems. Leaves up to 2 meters long, linear, flat or nearly terete, with sheathing base. Flowers naked, covered by hyaline envelopes, in 2 alternate rows in a 1-sided spike; spikes solitary or 2 or 3 within a spathe. Staminate flower a single sessile anther. Pistillate flower of 2 fused carpels; style short, with 2 filiform stigmas. Fruit 1-seeded, coriaceous, ovate, flattened, crowned by the style base, with two cordate or sagittate lobes at the base. Seed globose, not ridged. — A few species on marine shores of the Pacific Coast.

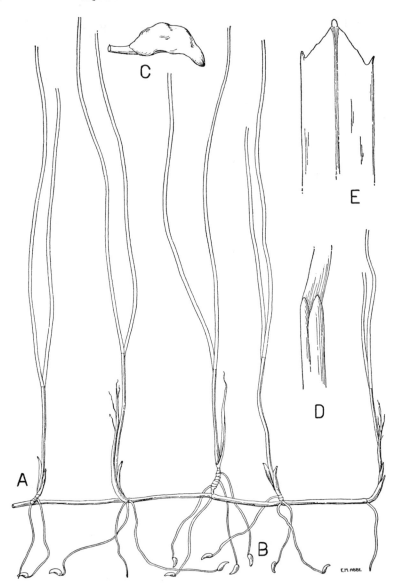

FIG. 22. *Halodule wrightii.*

(A) Sterile plant in winter condition showing habit with rhizome bearing short, erect branches with leaves; x 1. (B) Fleshy tuber-like storage organs on tips of roots; x 1. (C) Enlarged fleshy root; x 3. (D) Base of blade showing auricles at junction with sheath; x 3. (E) Apex of leaf; x 10.

KEY TO SPECIES OF PHYLLOSPADIX

1. Spathe with 1 (rarely 2) spikes, borne about 1 to 2 dm. above the base of the stem; leaves at least 2 mm. wide, flat .... *P. scouleri*
1. Spathe with 2 or 3 spikes, borne 4 to 10 dm. above the base of the stem; leaves not more than 1.5 mm. wide, often terete or complicate .... *P. torreyi*

Phyllospadix scouleri Hook. Fig. 24, E–H. Map 50.
Near the low-tide mark. On surf-beaten rocky shores along the Pacific Ocean.

Phyllospadix torreyi Wats. Map 50.
Usually below low-tide mark in quiet waters along the coast of central California.

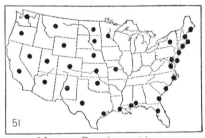

Map 50. *Phyllospadix scouleri*, ●;
P. *torreyi*, ▲.

Map 51. *Ruppia maritima*.

RUPPIA: Widgeon Grass, Ditch Grass
Submersed perennials with simple or much-branched stems. Leaves linear, with tapering apex and broad, sheathing base. Flowers perfect, naked, 2 to 4 in a cluster terminating a slender peduncle, wrapped in the sheathing leaf bases before opening. Stamens 2, sessile; carpels 4, sessile, becoming long-stipitate as they mature; stigma peltate or on the end of the style. Fruits oblique drupelets on long stipes, usually 4 in a cluster on a long peduncle which contracts and often becomes coiled. Seed with a coiled embryo. — Three or 4 widespread species.

Ruppia maritima L. Widgeon grass. Figs. 23, A–C; 25, G–H. Map 51.
Widespread, chiefly in brackish waters along the Atlantic Coast and in alkaline lakes, ponds, and streams in the western United States.

Note. *R. occidentalis* Wats. has been described from western North America. It appears to be a form of the preceding species differing chiefly in its longer stipular sheaths (15 mm. long) which are free for about ½ their length.

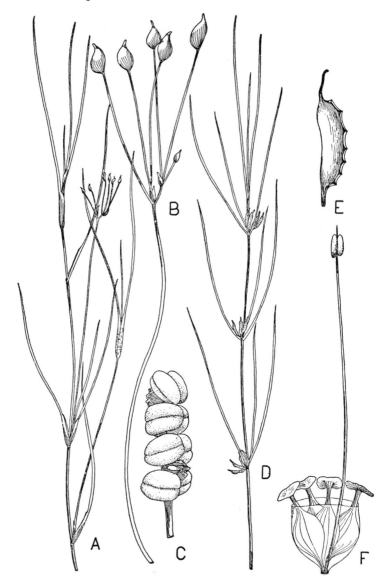

Fig. 23. *Ruppia maritima* (A–C) and *Zannichellia palustris* (D–F).

(A) Part of plant; x 1. (B) Cluster of fruits on long peduncle; x 5. (C) Peduncle with 2 flowers, each with 4 stamens and 4 carpels; x 10.

(D) Fruiting branch; x 1. (E) Fruit; x 10. (F) Flower showing 4 carpels and 1 stamen; x 10.

REFERENCES

Fernald, M. L., and K. M. Wiegand. The genus *Ruppia* in eastern North America. Rhodora 16:119–127. 1914.
Graves, A. H. The morphology of *Ruppia maritima*. Conn. Acad. Arts and Sci. Trans. 14:59–170. 1908.

ZANNICHELLIA: Horned Pondweed

Submersed monoecious perennials with slender, simple or branched, leafy stems. Leaves mostly opposite, linear, entire, with sheathing or free stipules. Flowers naked, axillary, in a hyaline, deciduous spathe; both kinds, 1 staminate and 2 to 5 pistillate, often in the same axillary, cup-shaped involucre; staminate flower a single stamen; pistillate flowers of solitary carpels, becoming stipitate. Fruit a stipitate, curved nutlet, toothed or entire along the margin. — One or 2 species.

Zannichellia palustris L. Fig. 23, D–F. Map 52.

In lakes, bays, streams, and ditches in fresh or brackish water. Widespread except in the extreme South.

Map 52. *Zannichellia palustris*.   Map 53. *Zostera marina*.

ZOSTERA: Eelgrass

Submersed monoecious perennials with creeping and branching rootstocks and branched, flattened, leafy stems. Leaves up to 2 meters long, alternate, 2-ranked, linear, flat, and sheathing at base. Flowers imperfect, the staminate and pistillate alternating, in 2 rows in a 1-sided spike borne solitary in a spathe. Perianth none. Staminate flower a single sessile anther. Pistillate flower of 2 fused carpels; style slender, terminating in 2 slender, deciduous stigmas. Fruit a narrowly ovoid utricle, rounded at base; seed with prominent ridges showing through the thin pericarp. — A few widespread species limited to marine waters in temperate regions.

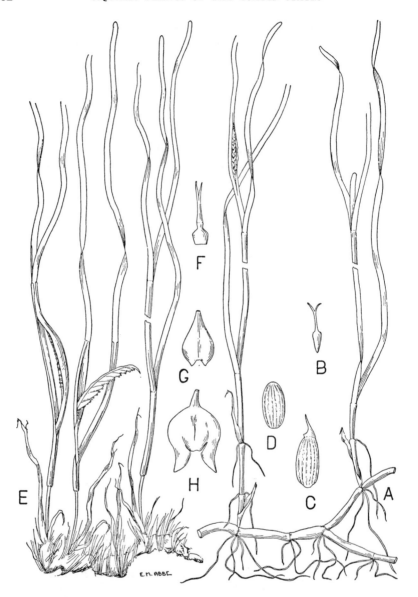

Fig. 24. *Zostera marina* (A-D) and *Phyllospadix scouleri* (E-H).

(A) Plant with flowering spadix; x 1/3. (B) Pistil; x 3. (C) Mature fruit; x 3. (D) Seed; x 3.

(E) Plant with creeping rhizome and spadix; x 1/3. (F) Pistil with 2-lobed stigma; x 3. (G) Pistil, later stage, after style is abscised; x 3. (H) Fruit showing basal lobes; x 3.

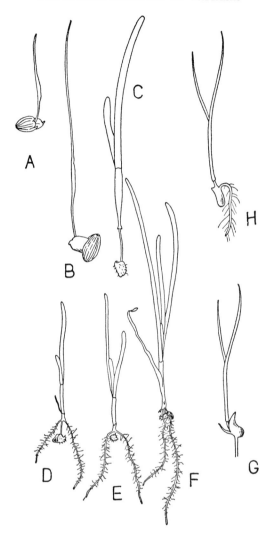

Fig. 25. Seedlings of *Zostera marina* (A–F) and *Ruppia maritima* (G, H).

(A, B) Early stages of seedlings; x 2. (C) Seedling with roots beginning at first node; x 1. (D–F) Later stages of seedlings with a pair of secondary roots; x 1/2. (A–F, from W. A. Setchell.)

(G) Seedling with pericarp wall attached to base; x 2. (H) Later stage of seedling showing root; x 2.

Zostera marina L. Eelgrass, Wrack. Figs. 24, A–D; 25, A–F; 48a, B. Map 53.

In shallow bays, mostly on soft bottoms in the intertidal zone; often the dominant vegetation over large areas. Frequently intermixed with marine algae some of which grow attached to it. This is by far the most widespread and abundant seed plant of salt water in temperate regions. Common along the Pacific Coast and formerly also abundant along the North Atlantic Coast southward to North Carolina.

NOTE. About 1930 *Z. marina* began to disappear at an alarming rate from the Atlantic Coast. The plants discolored, decayed, and gradually wasted away, first the leaves and stems and finally the rootstocks. Within a few years all the eelgrass had disappeared from the extensive areas over which it formerly was the dominant vegetation. Its disappearance created a serious shortage of feed for certain waterfowl such as geese and brant. Several industries utilizing the dried leaves of eelgrass or "wrack" for stuffing cushions and mattresses or for the manufacture of insulating or soundproofing materials had to discontinue operations entirely.

Various theories have been proposed to explain the cause of the so-called "wasting disease" of eelgrass. Some biologists maintain that the disease is produced by causal organisms. Among the organisms found more or less constantly associated with diseased plants under some conditions are a mycetozoan parasite, *Labyrinthula,* a fungus, *Ophiobolus,* and bacteria. Other biologists have explained the disappearance of the eelgrass as caused by changes in the environment. Among the causal agents mentioned are oil wastes in the water, changes in the salinity of the sea water, fluctuations in temperature, and changes in currents and phases of the moon. Probably both organisms and changes in environmental factors were concerned in producing this epidemic. Since 1938 the eelgrass has re-established itself, at least in small local colonies in the shallow bays about the eastern end of Long Island, New York.

REFERENCES

Cottam, C. The present situation regarding eel-grass. Trans. Amer. Game Conf. 21:295. 1935.

———. The eel-grass situation on the American Pacific Coast. Rhodora 41:257–260. 1939.

Mounce, I., and W. Diehl. A new *Ophiobolus* on eel-grass. Canadian Jour. Res. 11:242. 1934.

Stevens, N. E. Environmental conditions of the wasting disease of eel-grass. Science, n.s. 84:87–89. 1936.

Renn, C. E. Persistence of the eel-grass disease on the American Atlantic Coast. Nature 138:507–508. 1936.

———. The eel-grass situation along the Middle Atlantic Coast. Ecology 18:323–325. 1937.

Setchell, W. A. Morphological and phenological notes on *Zostera marina* L. Univ. Calif. Publ. in Botany 14:389–452. 1929.

Young, E. L. *Labyrinthula* on Pacific Coast Eel-grass. Canadian Jour. Res. Sect. C. 16:115–117. 1938.

———. Studies on *Labyrinthula*, the etiologic agent of the wasting disease of eel-grass. Amer. Jour. Bot. 30: 586–593. 1943.

NAJADACEAE: Naiad Family

Submersed aquatic herbs with slender branches and fibrous roots. Leaves opposite or crowded into apparent whorls, linear or linear-lanceolate, either spiny-toothed, finely serrate, or almost entire, dilated at base and often with prominent auricles or "stipules." Plants monoecious or dioecious; flowers imperfect, axillary, usually solitary and sessile. — A single genus with about 35 species inhabiting fresh or brackish waters of temperate and tropical regions.

NAJAS: Naiad

Staminate flowers consisting of a single, sessile, 1- or 4-celled anther, usually surrounded by 2 envelopes, one considered a perianth, the other a spathe. Pistillate flowers of a single carpel, surrounded by either 1 or 2 envelopes. Fruit an achene. — Eight species occur in the United States.

KEY TO SPECIES OF NAJAS

1. Leaves coarsely toothed (spines discernible without a lens), bright green; internodes and backs of leaves often spiny; seeds large, usually 4 to 5 mm. long, 2 to 3 mm. wide, finely reticulate; plants dioecious .... *N. marina*
1. Leaves almost entire or toothed (spines except in *N. minor* usually not discernible without a lens), often olive green or at times reddish; internodes and backs of leaves never spiny; seeds slenderer and smaller, 4 mm. long or less, 2 mm. wide or less; the seed coat variously reticulate; plants monoecious.
    2. Leaf bases broadly and truncately lobed or auriculate.
        3. Leaf bases broadly and truncately lobed; areolae of seed coat much broader than long, arranged in regular, vertical rows; leaves somewhat stiff, recurved, spiny .... *N. minor*
        3. Leaf bases auriculate and scarious, decidedly spiny-toothed; seeds very slender, often with a slight tendency to be curved; areolae longer than broad, sunken in, giving the seed coat a decidedly roughened appearance; leaves slender, not stiff, not recurved .... *N. gracillima*

2. Leaf bases neither broadly and truncately lobed nor auriculate, but little enlarged, and sloping.
    4. Seed coat smooth and glossy, finely reticulate with 30 to 50 rows of areolae (inconspicuous) around the seed; styles slender, 1 mm. or more long.
        5. Leaves finely and closely spined .... *N. flexilis*
        5. Leaves prominently spined .... *N. conferta*
    4. Seed coat smooth or pitted, not glossy; styles stout.
        6. Seed coat very finely reticulate with 50 to 60 longitudinal rows of areolae around the seed, these typically rectangular and sunken, giving the seed coat a roughened appearance; styles rather stout, 0.7 to 1.2 (1.5) mm. long; leaves finely serrate, with about 50 teeth on a margin .... *N. muenscheri*
        6. Seed coat coarsely reticulate with 10 to 30 rows of areolae around the seed, these smooth or sunken.
            7. Seed coat smooth; seeds 2.3 to 2.5 mm. long; shoots stout, 1 to 1.5 mm. thick .... *N. olivacea*
            7. Seed coat rough, shallowly pitted; seeds 2 mm. long; shoots slender and wiry, 1 mm. or less thick .... *N. guadalupensis*

Najas conferta A. Br. Fig. 28, H. Map 54.

In ponds and lakes. Florida.

Najas flexilis (Willd.) R. and S. Figs. 26, B; 29, E–H; 28, B. Map 55.

A variable species with long, slender stems in deep water and forming short, compact, bushy plants in shallow water along sandy shores. Common in shallow ponds, lakes, and sluggish streams. Widespread in the northern states.

MAP 54. *Najas conferta.*   MAP 55. *Najas flexilis.*

Najas gracillima (A. Br.) Morong. Figs. 26, C; 28, F. Map 56.

Local on sandy bottoms of lakes and ponds, mostly in acid water. Local in the northeastern states and infrequent in the north central states.

Najas guadalupensis (Spreng.) Morong. Figs. 26, A; 28, E. Map 57.

This variable species seldom fruits in the north. In the southern states forms occur with broader leaves. Common in the southern states, local in the northeastern and central states.

MAP 56. *Najas gracillima*.   MAP 57. *Najas guadalupensis*.

Najas marina L. Figs. 27, B; 29, A–D; 28, A. Map 58.

Widespread but very local in shallow, brackish ponds, lakes, and streams.

Najas minor Allioni. Figs. 27, A; 28, G. Map 59.

Locally common in the lower Hudson River, where it was first found in America. It has also been found along Lake Ontario and in the Tennessee Valley. Its range is but imperfectly known. Its recent discovery has suggested the possibility that it was introduced from Europe.

MAP 58. *Najas marina*.   MAP 59. *Najas minor*.

Najas muenscheri Clausen. Figs. 27, C; 28, C. Map 60.

Common on the tidal flats of the lower Hudson River.

Najas olivacea Rosen. and Butt. Figs. 27, D; 28, D. Map 61.

A coarse-stemmed species, mostly found in sterile state; fruits have been found but rarely. Very local in deep water in Cayuga Lake, New York, Minnesota, and Michigan.

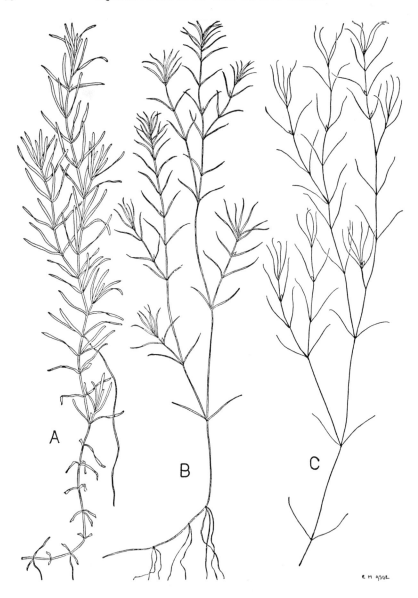

Fig. 26. *Najas guadalupensis* (A), *N. flexilis* (B), and *N. gracillima* (C). Habit sketches of plants; x 1/2.

Fig. 27. *Najas minor* (A), *N. marina* (B), *N. muenscheri* (C), and *N. olivacea* (D). Habit sketches of plants; x 1/2.

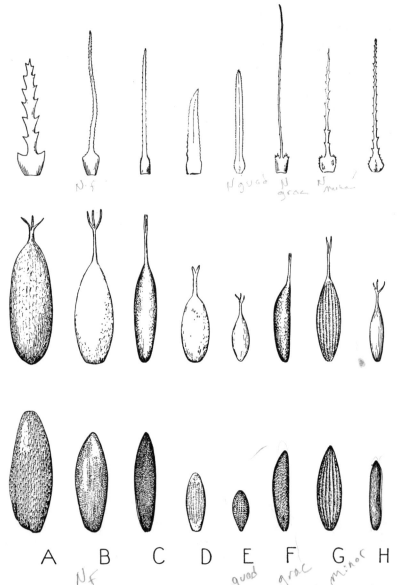

Fig. 28. *Najas marina* (A), *N. flexilis* (B), *N. muenscheri* (C), *N. olivacea* (D), *N. guadalupensis* (E), *N. gracillima* (F), *N. minor* (G), and *N. conferta* (H). Leaves (upper row), pistillate flowers (middle row), and seeds (lower row). Leaves, x 4. Flowers and seeds, x 16; except E and G, x 24. (From R. T. Clausen.)

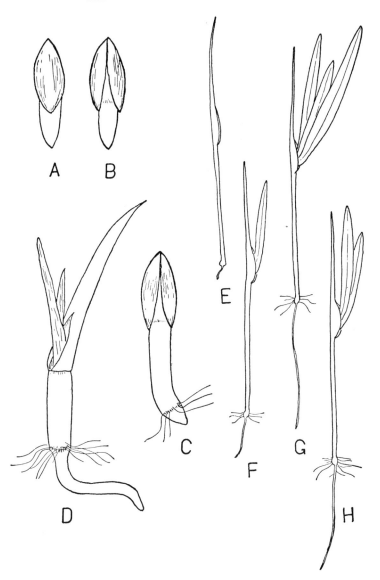

Fig. 29. *Najas marina* (A–D) and *N. flexilis* (E–H).

(A, B) Germinating seeds, 15 days after planting; x 6. (C) Embryo with seed coat on top of cotyledon and ring of root hairs on upper part of radicle; x 6. (D) Seedling 10 days old showing primary root, cotyledon, and two leaves from plumule; x 6.

(E–H) Successive stages in the development of seedlings; x 3.

MAP 60. *Najas muenscheri*.   MAP 61. *Najas olivacea*.

REFERENCES

Fernald, M. L. Notes on the distribution of *Najas* in northeastern America. Rhodora 25:105–109. 1923.

Clausen, R. T. Studies in the genus *Najas* in the northern United States. Rhodora 38:333–345. 1936.

———. A new species of *Najas* from the Hudson River. Rhodora 39:57–60. 1937.

Rosendahl, C. O., and F. K. Butters. The genus *Najas* in Minnesota. Rhodora 37: 345–348. 1935.

Rosendahl, C. O. Additional notes on *Najas* in Minnesota. Rhodora 41:187–189. 1939.

JUNCAGINACEAE: Arrow-Grass Family

Perennial or annual, subaquatic or marsh herbs with mostly basal, rushlike leaves with sheathing bases. Flowers in racemes or spikes or some of them axillary and basal, perfect or imperfect, naked or with a 3- or 6-parted perianth; carpels 3 to 6, or solitary; stamens 3, 6, or solitary. Fruit follicular, capsular, or indehiscent; seeds with straight embryo, without endosperm.

KEY TO GENERA

1. Leaves alternate on the erect stem; rootstock creeping; carpels 3 to 6, more or less fused; perianth persistent .... *Scheuchzeria*
1. Leaves all basal, from a short rootstock.
    2. Flowers in terminal spikes or racemes; carpels 3 to 6, more or less fused; perianth present .... *Triglochin*
    2. Flowers of 2 kinds, the basal solitary in the axils of leaves, others in terminal spikes; carpels solitary; perianth wanting .... *Lilaea*

LILAEA: Flowering Quillwort

Acaulescent annual herb with fibrous roots and clustered, basal, terete, sheathing leaves from 5 to 60 cm. long. Flowers perfect or imperfect, naked, most of them in many-flowered spikes terminating the scapes; a few of the flowers are basal in the axils of the sheathing leaf bases. The spicate flowers staminate, pistillate, or perfect, all intermixed; staminate flower a solitary, sessile stamen with a 2-celled anther and bractlike connective; pistillate flower a solitary carpel with short style, usually subtended by a bract; perfect flowers consisting of a stamen and a carpel. The basal flowers pistillate, composed of a solitary carpel with a filiform style 3 to 10 cm. long. Fruit of the spicate flowers a winged, ridged achene; fruit of the basal flowers wingless and larger. This monotypic genus has by some been placed in a separate family, Lilaeaceae.

Map 62. *Lilaea subulata.*   Map 63. *Scheuchzeria palustris.*

Lilaea subulata H. and B. Fig. 30. Map 62.

Local on tidal mud flats and marshes near the Pacific Coast; in mud of ponds on alkaline flats and meadows inland in California, Idaho, and Nevada. From British Columbia to South America.

Note. *Lilaea* has much the same general appearance as *Isoëtes* or young, sterile clumps of *Juncus*. Since it usually grows on rather inaccessible, muddy tidal flats near the river mouths where it is alternately inundated and exposed, it is easily overlooked. This may account for the apparent paucity of reported stations.

SCHEUCHZERIA

Perennial herbs with short, creeping, jointed rootstocks and erect, simple, leafy stems with remains of old leaves attached at base. Leaves linear to tubular, 10 to 40 cm. long, striated, with a membranous, ligulate sheath at base; upper leaves reduced to bracts. Flowers perfect, small, few, in lax, bracteate racemes; perianth of 3 sepals and 3 petals, greenish-yellow, persistent; stamens 6, inserted at the base of the perianth parts; carpels 3 (rarely 4 to 6), globular, distinct or fused at the

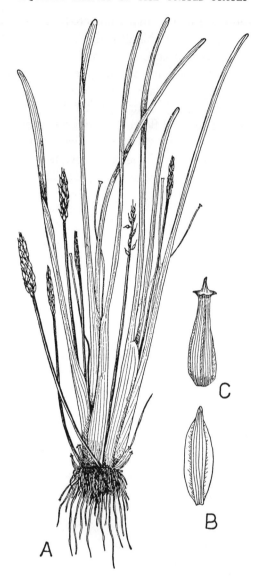

Fig. 30. *Lilaea subulata.*
(A) Plant; x 1/2. (B) Achene from spike; x 3. (C) Achene from basal flower; x 3.

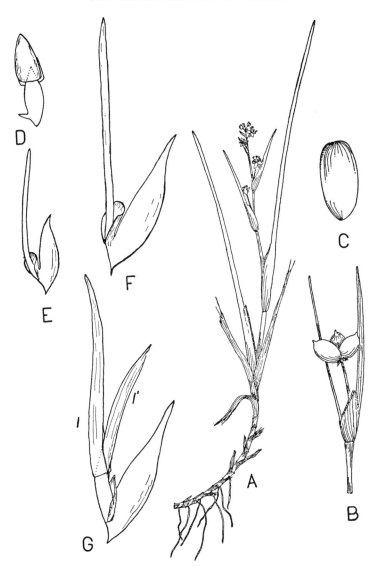

Fig. 31. *Scheuchzeria palustris.*

(A) Plant; x 1/2. (B) Fruit cluster; x 1. (C) Seed; x 6. (D) Germinating seed with embryo pushing out of seed coat, 15 days after planting; x 4. (E) Young seedling, 3 days after germination; x 4. (F) Seedling with elongated plumule, 6 days after germination; x 6. (G) Seedling with first and second leaves, 1 and 1'; x 6.

base, 1-celled; fruits follicle-like, inflated, with 1 or 2 seeds, dehiscent. Seeds straight or nearly so. — One circumpolar species.

Scheuchzeria palustris L. Fig. 31. Map 63.

Widespread but local, mostly in wet sphagnum bogs. Northeastern states and infrequent in the western states.

TRIGLOCHIN: Arrow Grass

Acaulescent, perennial herbs with short rootstocks with attached bases of old leaves; leaves clustered, basal, rushlike with ligulate, membranous sheaths. Flowers perfect, numerous in a slender, spikelike raceme terminating the scape; perianth of 3 or 6 small greenish segments; stamens 3 to 6, sessile or nearly so; carpels 3 or 6, fused into a compound pistil with sessile, plumose stigmas. Fruit a cluster of 3 to 6 one-seeded carpels separating at maturity from the base upward from a central axis and dehiscing; seeds compressed or angular. — Several widespread species of both hemispheres.

Note. The leaves of *Triglochin* sometimes cause poisoning when eaten by cattle or sheep. Their toxic properties appear to be due to the production of hydrocyanic acid.

KEY TO SPECIES OF TRIGLOCHIN

1. Plants stout, usually with several scapes; carpels 6 (rarely 3); fruit about twice as long as thick .... *T. maritima*
1. Plants slender, mostly with 1 or 2 scapes; carpels 3.
    2. Fruit linear, tapering at base, much longer than thick .... *T. palustris*
    2. Fruit globose or 3-lobed, thicker than high .... *T. striata*

Triglochin maritima L. Fig. 32, A–D. Map 64.

Common in brackish or salt marshes along the coasts; also widespread in low, wet meadows and marl bogs and on alkaline or brackish flats. Chiefly in the western states.

Triglochin palustris L. Fig. 32, E–G. Map 65.

Local in wet marl bogs and springy places. Chiefly in the northeastern states and rare in the Rocky Mountain states.

Triglochin striata R. and P. Fig. 32, H–I. Map 65.

Local in brackish and salt marshes. Along the South Atlantic Coast, also along the Gulf Coast and Pacific Coast.

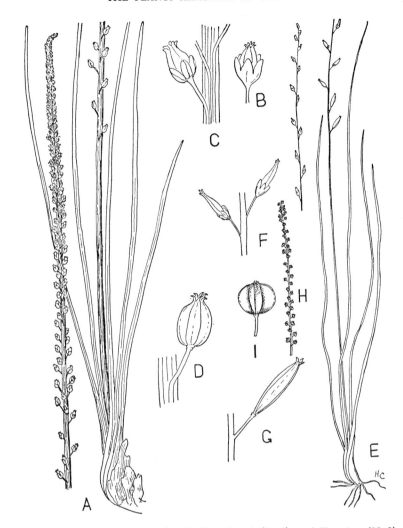

Fig. 32. *Triglochin maritima* (A–D), *T. palustris* (E–G), and *T. striata* (H, I).

(A) Plant; x 1/3. (B) Flower with perianth; x 3. (C) Flower with perianth removed to show stamens; x 3. (D) Mature capsule with six carpels; x 3.

(E) Plant; x 1/3. (F) Flowers; x 3. (G) Mature capsule with three carpels; x 3.

(H) Spike; x 1/3. (I) Capsule; x 3. (From Muenscher, 1939).

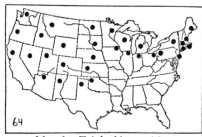

Map 64. *Triglochin maritima.*

Map 65. *Triglochin palustris,* ●; *T. striata,* ▲.

REFERENCES

Morong, Thomas. The Najadaceae of North America. Torrey Bot. Club Mem. 3:1–65. 1893.

Clausen, A. B., and E. A. Moran. Toxicity of arrowgrass for sheep and remedial treatment. U. S. Dept. Agr., Techn. Bull. 580. 1937.

Muenscher, W. C. *Lilaea subulata* in Washington. Torreya 38:8. 1938.

ALISMACEAE: Water-Plantain Family

Mostly aquatic or marsh perennials or, rarely, annuals, with sheathing leaves and scapelike stems from short, erect rootstocks, rhizomes, or tubers; roots fibrous. Leaves radical, with long petioles and emersed or floating blades, or in submersed forms subulate or consisting of phyllodia; blades often with a basal lobe on each side, sagittate or hastate. Inflorescence a raceme or panicle, mostly verticillate, emersed, rarely floating or submersed. Flowers perfect or imperfect, regular. Receptacle from flat to convex; sepals 3, separate, greenish, usually persistent; petals 3, separate, mostly white to pink or rose-colored, deciduous; stamens 6 or more, distinct; carpels from several to numerous, distinct, in a single whorl forming a disk or attached all over the receptacle and forming a head; ovary superior, 1-celled; fruit an achene with persistent style, flattened or turgid. — About 10 genera with about 50 species; widely distributed in marshes and shallow water.

KEY TO GENERA

1. Carpels attached on the receptacle in a ring; achenes forming a disk; flowers perfect.
    2. Style apical; achenes with long beak; petals dentate .... *Damasonium*
    2. Style lateral; achenes with minute beak; petals entire .... *Alisma*
1. Carpels attached over the entire surface of the receptacle; achenes forming a head.
    3. Upper and lower flowers of inflorescence perfect .... *Echinodorus*
    3. Upper flowers of inflorescence staminate; lower flowers pistillate or perfect.
        4. Lower flowers pistillate; achenes flattened .... *Sagittaria*
        4. Lower flowers perfect; achenes turgid .... *Lophotocarpus*

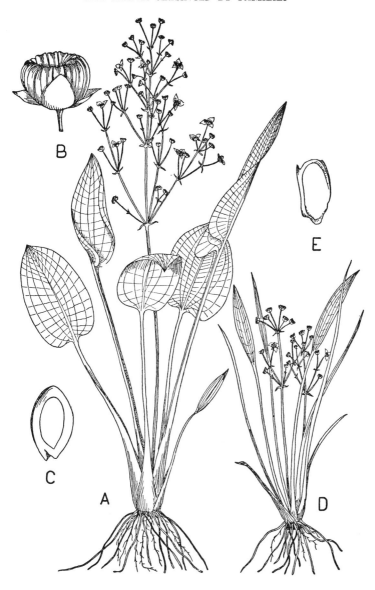

Fig. 33. *Alisma plantago-aquatica* (A–C) and *A. gramineum* (D, E).
(A) Plant; x 1/2. (B) Head of achenes; x 5. (C) Achene; x 10.
(D) Plant; x 1. (E) Achene; x 10.

ALISMA: WATER PLANTAIN

Perennial herbs with emersed or floating leaves from a short, erect, cormlike rootstock. Blades unlobed, cordate or tapering at base, oblong, ovate to lanceolate, or linear in outline. Flowers perfect, in verticillate, mostly compound panicles. Sepals 3, green; petals 3, white or pinkish; stamens 6, in pairs opposite the petals; carpels numerous, attached in a ring on the flattened or depressed receptacle. Achenes flattened on the sides and keeled on the back, with short lateral beak. — A few variable species widely distributed in the Northern Hemisphere.

KEY TO SPECIES OF ALISMA

1. Peduncles and pedicels erect; scape much longer than the leaves; achenes longer than wide, 2-ridged on back .... *A. plantago-aquatica*
1. Peduncles and pedicels recurved; scape rarely longer than the leaves; achenes as wide as long, 3-ridged on back .... *A. gramineum*

Alisma plantago-aquatica L. Water plantain. Figs. 33, A–C; 43, A. Map 66.

In shallow water and wet margins of lakes, temporary ponds, sloughs, and ditches. Common throughout the United States.

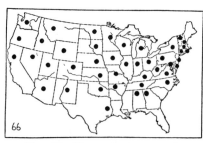 

MAP 66. *Alisma plantago-aquatica.*   MAP 67. *Alisma gramineum.*

Alisma gramineum Gmel. Narrow-leaved water plantain. Fig. 33, D–E. Map 67.

Locally common in shallow ponds and on mud along riverbanks; usually in alkaline soils or in limestone regions. Forms with fruiting scapes shorter than the leaves have been described as *A. gramineum* var. *geyeri* (Torr.) Samuelson. These occur nearly throughout the American range of the typical species.

FIG. 34. *Damasonium californicum* (A–C) and *Lophotocarpus calycinus* (D, E).
(A) Plant; x 1/4. (B) Head of achenes; x 1. (C) Achene; x 5.
(D) Plant; x 1/4. (E) Achene; x 10.

DAMASONIUM

Perennials with long-petioled leaves from a short, erect, cormlike rootstock. Leaves emersed or with lax petioles and floating blades; blades linear-lanceolate to ovate, 3-nerved. Flowers perfect, in a simple panicle of several verticils. Sepals 3, persistent; petals 3, white, with dentate margin; stamens 6, in pairs opposite the petals; carpels 6 to 14, in a whorl. Achenes in a disk, ribbed on the back and bearing an apical beak at least as long as the body. — A single species.

Damasonium californicum Torr. Fig. 34, A–C. Map 68.

Local in shallow water in ponds and borders of lakes in arid regions. From eastern Oregon to Nevada and northern California.

ECHINODORUS: Burheads

Annuals or perennials, mostly with runners; leaves radical on long petioles, or submersed and often reduced to phyllodia; blades lanceolate to ovate, with cordate base and several prominent veins. Flowers all perfect, on short pedicels, in verticils in open, elongated, compound or simple panicles; stamens 6 to 30; carpels numerous, on a convex receptacle, in fruit forming a bristly head; achenes with apical beak. Plants with much the same general aspect as *Sagittaria*. — Mostly limited to warmer regions; a few species occur in the United States.

KEY TO SPECIES OF ECHINODORUS

1. Scape up to 1 dm. long; leaves linear to lanceolate; stamens 9 .... *E. tenellus*
1. Scape 2 to 12 dm. long; leaves mostly broadly ovate to cordate; stamens 12 to 21.
   2. Scape erect; stamens 12 .... *E. cordifolius*
   2. Scape prostrate or rooted at nodes; stamens about 20 .... *E. radicans*

Echinodorus cordifolius (L.) Griseb. Fig. 35, A–B. Map 69.

In shallow ponds and borders of streams, ditches, and sloughs. Mostly in the lower Mississippi Valley.

Map 68. *Damasonium californicum.*

Map 69. *Echinodorus cordifolius.*

THE PLANTS ARRANGED BY FAMILIES 83

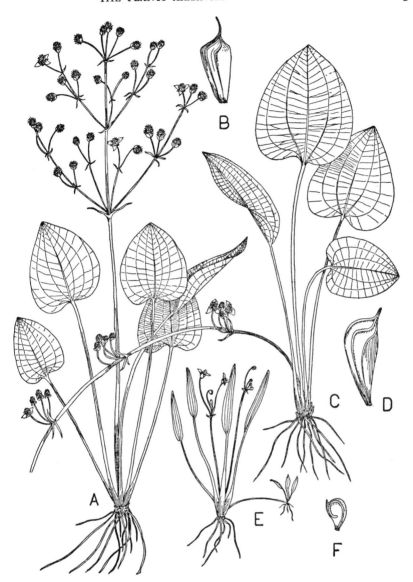

Fig. 35. *Echinodorus cordifolius* (A, B), *E. radicans* (C, D), and *E. tenellus* (E, F).

(A) Plant; x 1/2. (B) Achene; x 8.
(C) Plant; x 1/4. (D) Achene; x 8.
(E) Plant; x 1. (F) Achene; x 8.

Map 70. *Echinodorus radicans*.   Map 71. *Echinodorus tenellus*.

Echinodorus radicans (Nutt.) Engelm. Fig. 35, C–D. Map 70.

In ponds and low, wet places. Mostly in the Mississippi Valley and along the Atlantic Coast.

Echinodorus tenellus (Mart.) Buchenau. Fig. 35, E–F. Map 71.

Local, usually submersed on tidal mud or along stream banks. Mostly along the Atlantic Coast, rare inland.

LOPHOTOCARPUS

Annuals or perennials with much the same general habit and appearance as *Sagittaria*. Leaves with long, erect, somewhat fleshy petioles; blades lanceolate or broad and hastate at base; sometimes the leaves are reduced to fleshy, somewhat spongy phyllodia. Flowers in 2 to 5 whorls near the top of the scape, the upper staminate, the lower perfect. Sepals strongly concave; stamens 7 to 15; carpels numerous on a convex receptacle. Peduncles or scapes often recurved in fruit. Achenes crested or winged, more or less surrounded by the calyx. — About 6 species, mostly in warmer regions.

KEY TO SPECIES OF LOPHOTOCARPUS

1. Scapes shorter than the leaves; achene with a thin wing on back ....*L. calycinus*
1. Scapes about as long as the leaves; achene with a thick wing on back ....
....*L. californicus*

Lophotocarpus californicus J.G.Smith. Map 72.

Local in shallow water and on muddy shores of lakes and ponds. Oregon and California.

Lophotocarpus calycinus (Engelm.) J.G.Smith. Fig. 34, D–E. Map 73.

Local on tidal mud flats in brackish bays and estuaries along the Atlantic Coast; inland on muddy banks of streams, chiefly in the Mississippi Valley.

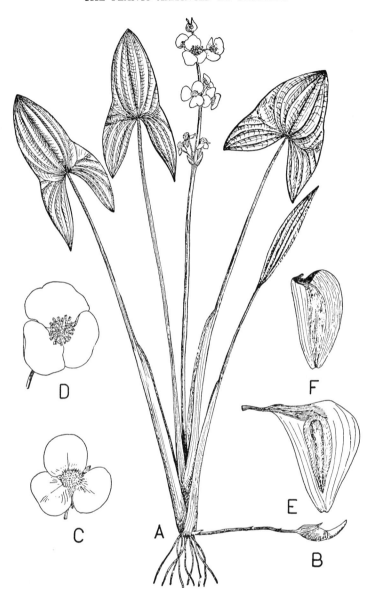

FIG. 36. *Sagittaria latifolia* (A–E) and *S. cuneata* (F).

(A) Plant showing pistillate flowers in the 2 lower whorls and staminate flowers above; x 1/3. (B) Young tuber. (C) Pistillate flower; x 1. (D) Staminate flower; x 1. (E) Achene; x 10.
(F) Achene; x 10.

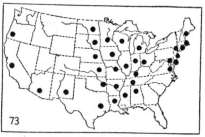

Map 72. *Lophotocarpus californicus.*   Map 73. *Lophotocarpus calycinus.*

Note. The brackish-water form, described as *L. calycinus* var. *spongiosus* (Engelm.) Fassett, has the leaves reduced to thick phyllodia or without basal lobes or distinct blade. The plants of the interior states are more robust and have larger, sagittate leaves.

SAGITTARIA: Arrowhead

Perennials with fleshy or tuber-bearing rootstocks and rosettes of sheathing, basal leaves. Leaves variable, subulate, and submersed or long-petioled with emersed or floating blades. Blades variable with the species and the depth of water, linear-lanceolate to lanceolate or elliptic; in the emersed state of many species, sagittate or hastate (Fig. 37). Flowers mostly in whorls of three in a simple inflorescence, those of the lower whorls pistillate or perfect and those of the upper whorls staminate. Sometimes the plants are dioecious. Staminate flowers with several to many separate stamens; pistillate flowers with numerous separate carpels on a convex to globose receptacle; achenes numerous, beaked, flattened, crowded in a globose head. — Several species, mostly of shallow water, muddy shores, and marshes.

Note. The tubers of several species are rich in starch and formed an important source of food for some of the Indian tribes.

KEY TO SPECIES OF SAGITTARIA

1. Fruiting heads with appressed sepals .... *S. montevidensis*
1. Fruiting heads with recurved sepals.
    2. Pedicels recurved when fruiting, thickened.
        3. Stamens with glabrous filaments; beak of achene nearly erect .... *S. subulata*
        3. Stamens with hairy filaments; beak of achene nearly horizontal .... *S. platyphylla*
    2. Pedicels not recurved when fruiting, slenderer.
        4. Beak of achene horizontal or nearly so (at right angles to the long axis).
            5. Beak long, slender; leaves sagittate .... *S. latifolia*
            5. Beak minute, short (mostly less than ¼ as long as the body of the achene).
                6. Stamen filaments dilated, short; leaves flat, mostly linear, rarely sagittate.

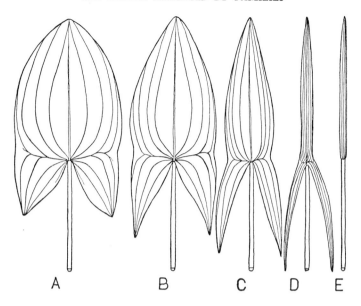

A B C D E

Fig. 37. Leaves of *Sagittaria latifolia* showing several types of blades.

       7. Anthers about as long as filaments .... *S. graminea*
       7. Anthers much longer than filaments .... *S. eatoni*
     6. Stamen filaments slender, not dilated; leaves often sagittate.
       8. Bracts of inflorescence lance-attenuate; pedicels of pistillate flowers long, slender.
         9. Bracts papillose; filaments glabrous .... *S. ambigua*
         9. Bracts smooth; filaments hairy .... *S. lancifolia*
       8. Bracts of inflorescence short, not attenuate.
         10. Bracts papillose; pedicels of pistillate flowers short .... *S. falcata*
         10. Bracts smooth; pedicels of pistillate flowers long, slender ......
           .... *S. weatherbiana*
  4. Beak of achene erect or nearly so.
    11. Beak long (about $\frac{1}{2}$ as long as the body of the achene).
      12. Leaves lanceolate to elliptical, rarely with short basal lobes; peduncle bent near lower whorl of fruiting pedicels .... *S. rigida*
      12. Leaves sagittate; peduncle straight.
        13. Blade lanceolate to linear, with linear basal lobes .. *S. engelmanniana*
        13. Blade ovate-oblong, with broader basal lobes .... *S. longirostra*
    11. Beak short (less than $\frac{1}{3}$ as long as the body of the achene).
      14. Leaves linear, rarely with blade, not sagittate; plant mostly submersed; flowers small .... *S. teres*
      14. Leaves sagittate; plant mostly emersed; flowers showy.
        15. Lowest bracts not over 1.5 cm. long; achenes with thick, nearly equal wings .... *S. cuneata*
        15. Lowest bracts at least 2 cm. long; achenes with thin, unequal wings .... *S. brevirostra*

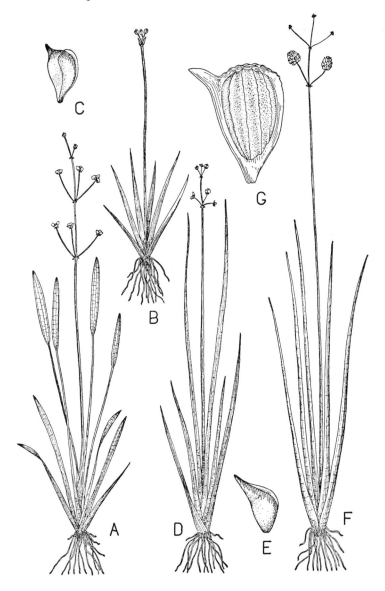

Fig. 38. *Sagittaria graminea* (A–C), *S. eatoni* (D, E), and *S. teres* (F, G).

(A) Plant from shallow water showing two kinds of leaves; x 1/2. (B) Small plant from deeper water showing a rosette of short, stiff, sharp-pointed leaves; x 1/2. (C) Achene; x 10.

(D) Plant; x 1/2. (E) Achene; x 10.
(F) Plant; x 1/2. (G) Achene; x 10.

Sagittaria ambigua J.G.Smith. Map 74.
Local in marshes, springy pools, and borders of streams. Kansas and Oklahoma.

Sagittaria brevirostra Mack. and Bush. Fig. 39, C–D. Map 74.
Local in sloughs and along riverbanks subject to inundation. Central Mississippi Valley.

Sagittaria cuneata Sheldon. Fig. 36, F. Map 75.
In shallow ponds, lakes, and backwaters along streams. Across the northern United States.

Sagittaria eatoni J.G.Smith. Fig. 38, D–E. Map 76.
Local in shallow ponds and backwaters of streams. North and Middle Atlantic Coast.

Sagittaria engelmanniana J.G.Smith. Fig. 40, C–D. Map 77.
In acid bogs and shallow ponds and lakes. Near the coast from Massachusetts to South Carolina.

Sagittaria falcata Pursh.
Local in ponds and on shores of lakes and streams. Delaware to Florida and Texas.

Sagittaria graminea Michx. Fig. 38, A–C. Map 78.
Common mostly on sandy bottom of shallow ponds and on margins of lakes and slow streams. Throughout the eastern United States. A variable species as regards leaf form and general habit. Several of the extremes have been treated as separate species.

Sagittaria lancifolia L. Fig. 42, E. Map 79.
Local in swamps and shallow ponds and along streams. Florida and the Gulf Coast.

Sagittaria latifolia Willd. Wapato, Duck potato. Figs. 36, A–E; 43, B; 90, A. Map 80.
Emersed in shallow water of ponds, lakes, and borders of slow streams, sometimes in acid bogs. Widespread throughout most of the United States except in the Southwest. In many places it has been introduced by the planting of its tubers to improve feeding places for waterfowl.

NOTE. This is a robust species with variable leaves. Specimens with hairy leaves have been described as var. *pubescens*. Extremes of leaf form have been designated as species, or by more conservative taxonomists as forms. The most common of these are: forma *obtusa,* with

Fig. 39. *Sagittaria longirostra* (A, B) and *S. brevirostra* (C, D).
(A) Plant; x 1/4. (B) Achene; x 10.
(C) Inflorescence; x 1/2. (D) Achene; x 10.

Fig. 40. *Sagittaria rigida* (A, B) and *S. engelmanniana* (C, D).
(A) Plant; x 1/2. (B) Achene; x 10.
(C) Plant; x 1/2. (D) Achene; x 10.

broad, obtuse blades; forma *gracilis,* with slender blades with narrow basal lobes; forma *hastata,* with broad blades with oblong, lanceolate, acute lobes. Plants with these leaf forms and others with intermediate leaves are frequently found growing in the same colony extending from shore into deep water, the plants in the deeper water producing the slender leaves (Fig. 37).

Sagittaria longirostra (Michx.) J.G.Smith. Fig. 39, A–B. Map 81.

In shallow water and springy bogs. Mostly in the southeastern states.

Map 74. *Sagittaria ambigua,* ●; *S. brevirostra,* ▲.

Map 75. *Sagittaria cuneata.*

Map 76. *Sagittaria eatoni.*

Map 77. *Sagittaria engelmanniana.*

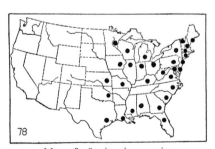

Map 78. *Sagittaria graminea.*

Map 79. *Sagittaria lancifolia.*

MAP 80. *Sagittaria latifolia.*

MAP 81. *Sagittaria longirostra.*

Sagittaria montevidensis C. and S.

In shallow ponds and slow streams. Naturalized from Brazil; local in the southeastern states and California. Grown in aquaria and summer pools northward.

Sagittaria platyphylla (Engelm.) J.G.Smith. Fig. 42, A–B. Map 82.

In sloughs and backwaters along rivers. Lower Mississippi Valley.

Sagittaria rigida Pursh. Fig. 40, A–B. Map 83.

Emersed in shallow to deep water in nonacid, muddy bottom, reaching its best development in coves and backwaters of streams and in bays of larger lakes. New England to Minnesota southward to Kansas and Virginia.

NOTE. The leaves of this species are extremely variable. Deep-water plants frequently produce only or mostly phyllodia; shallow-water plants produce leaves with broad blades and even basal lobes; intermediate plants may produce all types of leaves (Fig. 41).

Sagittaria subulata (L.) Buchenau. Fig. 42, C–D. Map 84.

Common in brackish tidal mud or silt, submersed or exposed during low tide, also in fresh-water ponds and streams. Chiefly along the Atlantic Coast, from Maine to Florida.

NOTE. The habit and size of the plants of this species vary with the depth of the water, the swiftness of the current, and the locality. Robust plants from Florida to Virginia with long ribbon-like, often floating leaves with dilated blades appear to be distinct from the typical forms, var. *typica,* with submersed, rather firm, linear leaves with blunt apex. The former have been designated as var. *natans* or var. *lorata.* Plants with very elongate, slender, submersed leaves, in running water of deep streams from Massachusetts to Pennsylvania, have been named var. *gracillima.*

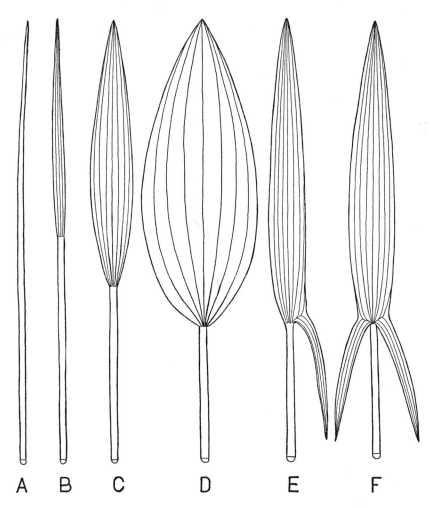

Fig. 41. Leaves from a single plant of *Sagittaria rigida* showing variation in the shape of the blades.

Sagittaria teres Wats. Fig. 38, F–G. Map 85.

Mostly submersed in shallow ponds. Near the coast from Massachusetts to New Jersey.

Sagittaria weatherbiana Fernald. Map 85.

In sloughs and backwaters along rivers. Along the Middle Atlantic Coast.

MAP 82. *Sagittaria platyphylla*.

MAP 83. *Sagittaria rigida*.

MAP 84. *Sagittaria subulata*.

MAP 85. *Sagittaria teres*, ●;
S. *weatherbiana*, ▲.

REFERENCES

Smith, J. G. North American species of *Sagittaria* and *Lophotocarpus*. Mo. Bot. Gard. Ann. Rept. 6:1-38. 29 pl. 1894.
Small, J. K. Alismaceae. In, North Amer. Flora. 17(1):42-62. 1909.
Fernald, M. L. Some forms in the Alismaceae. Rhodora 38:73-74. 1936.
———. *Sagittaria subulata*. Rhodora 42:408-409. 1940.
Clausen, R. T. The variations of *Sagittaria subulata*. Torreya 41:161-162. 1941.
Brown, W. F. A note on *Sagittaria kurziana*. Rhodora 44:211-213. 1942.
Buchenau, F. Alismataceae. In, Das Pflanzenreich, by A. Engler. 16(IV. 15):1-66. 1903.

BUTOMACEAE: FLOWERING-RUSH FAMILY

Perennial aquatic or marsh plants with short or long rootstocks. Leaves cauline or basal, sometimes with a milky sap. Flowers in umbels or axillary; sepals 3, persistent; petals 3, showy; stamens 9 to numerous, free, the external sometimes sterile; carpels 6 to 8, free or fused at the base, with numerous ovules attached all over the inner surface; fruits many-seeded, follicle-like, dehiscing on the inner side. Seeds without endosperm. — About 4 genera, mostly of warmer regions; representatives of 3 genera have been introduced into the United States.

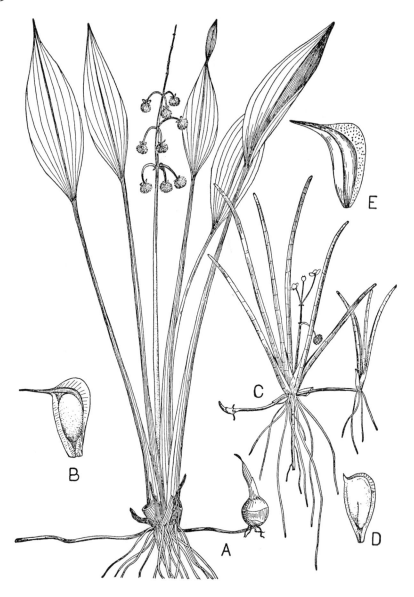

FIG. 42. *Sagittaria platyphylla* (A, B), *S. subulata* (C, D), and *S. lancifolia* (E).
(A) Plant; x 1/2. (B) Achene; x 10.
(C) Plant; x 1/2. (D) Achene; x 10.
(E) Achene; x 10.

THE PLANTS ARRANGED BY FAMILIES 97

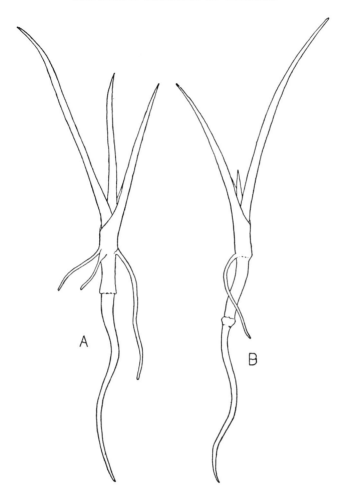

Fig. 43. Seedlings of *Alisma plantago-aquatica* (A) and *Sagittaria latifolia* (B) 10 days after germination in water; x 3.

KEY TO GENERA

1. Leaves all basal, sheathing, narrow, tapering; flowers in umbels, rose-colored .. ..*Butomus*
1. Leaves ovate, elliptic to cordate, petioled; flowers yellow or green.
    2. Rootstock short and thick; flowers in umbels; carpels 15 to 20, fused at base ....*Limnocharis*
    2. Rootstock slender, creeping, and branched; flowers mostly solitary in leaf axils; carpels 5 to 8, free ....*Hydrocleis*

BUTOMUS: Flowering Rush

Slender emersed plants with basal, 2-ranked leaves on short rhizomes bearing bulbils in their axils. Leaves sheathing and keeled at base, gradually narrowing into a slender, twisted, tapering blade about 1 meter long, erect or floating in deeper water. Flowers perfect, on slender pedicels arranged in a simple umbel terminating the erect, terete scape; scape about 6 to 12 dm. long. Sepals and petals each 3, elliptic, rose-colored, persistent; stamens 9; carpels 6, fused at base, each with many ovules attached all over the inner surface; styles apical, persistent, stigmatic on the inner side. Fruit of 6 coriaceous, follicle-like, many-seeded carpels dehiscing on the inner side.

Butomus umbellatus L. Fig. 44. Map 86.

Locally common in shallow to deep water or along wet shores where the water has receded, chiefly on muddy or silty bottom. Native to Eurasia. Introduced and spreading rapidly, especially in the Lake Champlain Basin and about the western end of Lake Erie. *Butomus* first became naturalized in North America in the marshes of the St. Lawrence River near Montreal.

Map 86. *Butomus umbellatus.*

Map 87. *Hydrocleis nymphoides.*

HYDROCLEIS

Perennials with creeping stems rooting at the nodes. Leaves alternate, petioled, with broadly-ovate to cordate, entire, glossy blade with spongy midrib on the under side, mostly floating. Flowers perfect, showy, axillary on long peduncles, raised well above the water and lasting but 1 day. Sepals 3, persistent; petals 3, light yellow, obovate, about 2 to 3 cm. long, fugacious; stamens numerous, the external sterile; carpels 5 to 8, free, gradually tapering into the styles. Seeds several. — About 3 species native to Brazil.

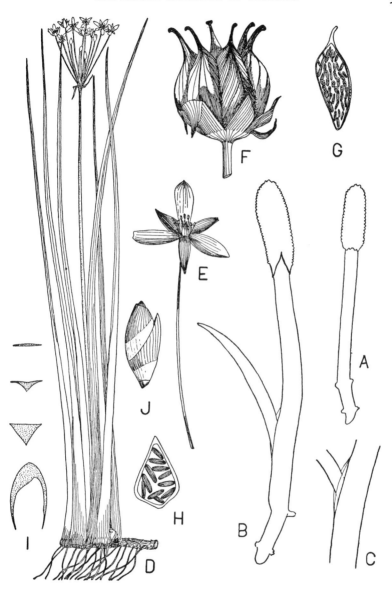

FIG. 44. *Butomus umbellatus*.

(A) Germinating seed; x 8. (B) Seedling; x 10. (C) Base of cotyledon; x 12. (D) Plant; x 1/6. (E) Flower; x 1/2. (F) Fruit; x 3. (G) Vertical section of a ripe carpel showing seeds attached all over the inner surface; x 3. (H) Diagram of cross section of carpel; x 5. (I) Cross section of leaf at base, near middle, and near apex; x 1. (J) "Bulbil" from leaf axil; x 2.

Hydrocleis nymphoides Buchenau. Water poppy. Fig. 45. Map 87.
Introduced in ponds and pools in the southern states.

NOTE. In the North it is a common ornamental plant in aquaria and summer pools. It grows well in shallow water but it will not survive the winter in the open.

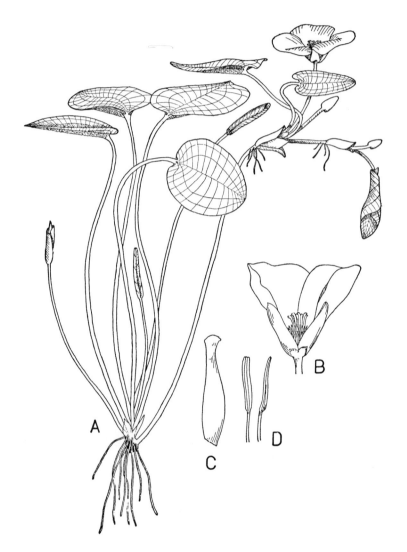

FIG. 45. *Hydrocleis nymphoides.*

(A) Habit sketch of plant from shallow water; x 1/2. (B) Flower; x 1. (C) A single carpel; x 3. (D) Stamens; x 3.

LIMNOCHARIS

Perennial, aquatic, stoloniferous herbs with erect, angled branches. Leaves emersed, alternate, lanceolate to broadly ovate, with angular petioles. Flowers in umbel-like clusters on 3-angled peduncles, perfect; sepals 3, persistent; petals 3, nearly round, pale yellow or with white margin, fugacious; stamens numerous, the outer sterile; carpels 15 to 20, in whorls, somewhat fused at base; stigmas sessile; fruiting scapes recurve into the water and develop new shoots.

Limnocharis flava (L.) Buchenau. Map 88.

Introduced from the West Indies or South America. Grown in aquaria, summer pools, and artificial ponds. Not winter hardy where frosts occur.

Map 88. *Limnocharis flava.*

Map 89. *Anacharis densa.*

REFERENCES

Core, E. L. *Butomus umbellatus* in America. Ohio Jour. Sci. 41:79–85. 1941.
Muenscher, W. C. *Butomus umbellatus* in the Lake Champlain Basin. Rhodora 32:19–20. 1930.

HYDROCHARITACEAE: Frogbit Family

Dioecious or, rarely, monoecious or polygamous, aquatic herbs, mostly with clustered leaves at nodes of rhizomes or a few with leafy stems. Flowers regular, sessile or on scapelike peduncles in a spathe; calyx usually of 3 sepals or lobes; corolla of 3 petals or wanting. In pistillate flowers the perianth fused into a tube and fused with the ovary of the compound pistil. Staminate flowers with 3 to 21 distinct or fused stamens. Pistillate flowers with 3 to 12 stigmas; ovary inferior, ripening under water into a few- or many-seeded, indehiscent fruit. — About 7 genera with widely distributed species of fresh or salt waters.

KEY TO GENERA

1. Leaves whorled, stems floating ....*Anacharis*
1. Leaves not whorled; stem forming a horizontal rootstock.
    2. Leaves opposite, on short, erect branches, petioled ....*Halophila*
    2. Leaves alternate, basal or at the nodes of the rootstock.
        3. Leaves petioled; blades rounded, spongy beneath; plant usually floating .. ..*Limnobium*
        3. Leaves sessile, linear (tapelike); plants anchored on the bottom.
            4. Branches of rootstock covered by persisting leaf bases; apex of leaf blunt, with sharp minute serrations ....*Thalassia*
            4. Branches of rootstock free from persisting leaf bases; apex of leaf tapering, without serrations ....*Vallisneria*

ANACHARIS (*ELODEA*): Waterweed

Perennial, slender-stemmed, branching, submersed aquatics with whorled or, rarely, opposite leaves and fibrous roots. Plants rooted on the bottom or sometimes forming extensive floating mats. Leaves from linear to oval-oblong or lanceolate, sessile, entire or finely serrate, thin. Plants dioecious or apparently polygamous. Flowers 1 to 3 in axillary, tubular, 2-cleft spathes. Staminate flowers with 3 sepals separate nearly to the base; petals 3, white or pinkish; stamens 3 to 9; the flowers break loose or extend to the surface to shed their pollen. Pistillate flowers with perianth tube with 6-parted limb, the tube elongated, surrounding the capillary style, and raising the 3 two-lobed stigmas to the surface. Ovary 1-celled, with 3 parietal placentae. Fruit coriaceous, indehiscent, oblong or spindle-shaped, about 8 mm. long and producing 1 to 5 spindle-shaped seeds about 4 mm. long (Fig. 47, E).

KEY TO SPECIES OF ANACHARIS

1. Flowers usually 3 in a spathe; leaves mostly 3 to 5 mm. wide, up to 4 cm. long; staminate flower about 10 to 12 mm. long, remaining attached by a long peduncle ....*A. densa*
1. Flowers solitary in the spathe.
    2. Leaves 1.2 to 4 mm. wide (average 2 mm.); spathe of staminate flower 10 to 13 mm. long, constricted at base into a stalk; staminate flower about 4 to 5 mm. long, remaining attached by a long peduncle; pistillate flowers with sepals about 2.5 mm. long ....*A. canadensis*
    2. Leaves 0.7 to 1.8 mm. wide (average 1.3 mm.); spathe of staminate flower globose, about 2 to 3 mm. long; staminate flower about 2 to 2.5 mm. long, sessile, breaking free from the spathe and rising to the surface before anthesis; pistillate flowers with sepals 1 to 1.8 mm. long ....*A. occidentalis*

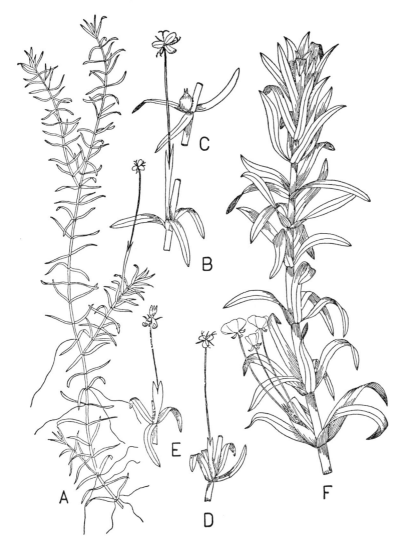

Fig. 46. *Anacharis occidentalis* (A–C), *A. canadensis* (D, E), and *A. densa* (F).
(A) Habit sketch of plant; x 1. (B) Pistillate flower; x 2. (C) Staminate flower; x 2.
(D) Pistillate flower; x 2. (E) Staminate flower; x 2.
(F) Section of plant with spathe containing 3 flowers; x 1.

Anacharis canadensis (Michx.) Planchon. Waterweed, Elodea. Fig. 46, D–E. Map 90.

Common in lakes, ponds, and slow streams, chiefly in calcareous waters. Throughout the northeastern and middle western states, local westward; also introduced in some places.

NOTE. This is the most common and widespread *Anacharis* in the United States. It is the "Elodea" most commonly used in biological laboratory experiments. It was early introduced, in its staminate form, into Europe where it soon escaped and became a pest in waterways. The staminate plant was by some considered to be a separate species, the *E. planchonii* Caspary.

Anacharis densa (Planchon) Marie-Vict. Brazilian waterweed. Fig. 46, F. Map 89.

Introduced from South America; naturalized in lakes and streams along the Atlantic Coast, in the Tennessee Valley, California, and probably elsewhere. This is the large "Elodea" in the trade. It is used in fish bowls, aquaria, and outdoor pools. The absence of earlier records suggests that it has become naturalized only in recent years.

Anacharis occidentalis (Pursh) Marie-Vict. Figs. 46, A–C; 47. Map 91.

On sandy bottom in lakes and streams, also on tidal flats of rivers; mostly in noncalcareous waters. Common in the northeastern and midwestern states.

NOTE. This species fruits abundantly on the tidal flats of the lower Hudson River. The fruits produce from 1 to 5 seeds. Of 100 fruits taken at random, 5 per cent contained 5 seeds each; 26 per cent 4 seeds; 38 per cent 3 seeds; 20 per cent 2 seeds, and 11 per cent 1 seed.

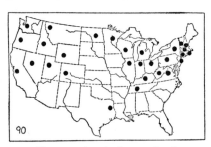 

MAP 90. *Anacharis canadensis*.   MAP 91. *Anacharis occidentalis*.

HALOPHILA

Submersed monoecious or dioecious perennials with creeping, scaly rootstocks, fibrous roots, and short, erect, leafy branches, each usually

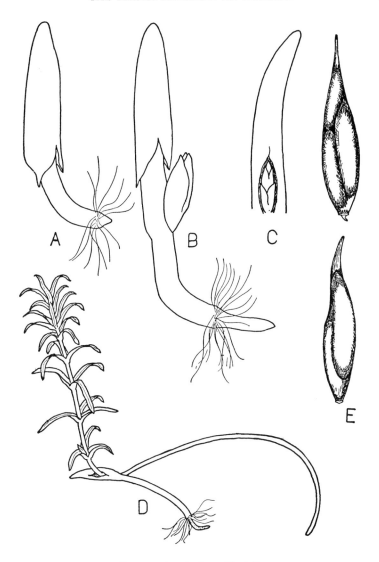

Fig. 47. *Anacharis occidentalis*.

(A) Germinating seed; x 8. (B) Seedling showing hypocotyl and plumule; x 8. (C) Front view of plumule; x 8. (D) Seedling about 10 days after seed germination; x 2. (E) Fruits containing several seeds; about x 4.

bearing a pair of opposite scales near the middle and 1 to 3 pairs of opposite leaves near the apex. Blades minutely serrate, oblong to linear-oblong. Flowers 1 or 2 borne in a 2-bracted sheath. Staminate flowers peduncled; perianth lobes 3; stamens 3, sessile, with filiform pollen. Pistillate flowers sessile, usually in the same sheath with the staminate; perianth lobes 3, minute; pistil compound, with 3 stigmas; ovary 1-celled, with 3 parietal placentae; fruit an ovoid, membranous capsule, many-seeded, inclosed in the sheath. — About 6 species widespread in marine waters of tropical regions.

KEY TO SPECIES OF HALOPHILA

1. Leaves 1 pair at the apex of branch, with slender petiole .... *H. baillonis*
1. Leaves 2 or 3 pairs near the apex of branch, sessile or with short, thick petiole
.... *H. engelmanni*

Halophila baillonis Aschers. Map 92.

Local on sandy or marly bottoms on reefs about the Florida Keys; more common in the West Indies.

Halophila engelmanni Aschers. Fig. 49, B–C. Map 92.

In salt water in shallow bays off the coast of southern Florida and Texas.

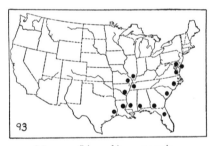

Map 92. *Halophila engelmanni*, ●; *H. baillonis*, ▲.

Map 93. *Limnobium spongia*.

LIMNOBIUM

Monoecious aquatics with rosettes of leaves and fibrous roots at the nodes of rhizomes. Leaves petioled, with ovate, cordate, or reniform entire blades with spongy lower surface. Staminate flowers 3 to 10 in a 2-bracted spathe on long peduncles; sepals 3, oblong; petals 3, linear; stamens 6 to 12, with filaments fused into a solid column; anthers unequal in length. Pistillate flowers 1 or 2 in a 2-bracted spathe on a short peduncle; sepals and petals 3; stamens of 3 to 6 subulate rudiments; ovary 6- to 9-celled; fruit a many-seeded, ovoid berry on a stout, recurved peduncle.

Limnobium spongia (Bosc.) Steud. Frogbit. Fig. 49, A. Map 93.

Floating in shallow stagnant water or rooted in mud. Lowlands along the South Atlantic Coast and the lower Mississippi Valley; infrequent elsewhere.

THALASSIA

Dioecious perennials with thick, crisp, creeping rootstocks and clusters of submersed leaves on short, erect branches. Leaves 2-ranked, sheathing at base, ribbon-like, 2 to 4 dm. long, about 1 cm. wide, with minutely serrate, rounded apex. The leaf bases of old leaves persist as fibrous shreds. Scapes from axils of leaf clusters, with solitary flowers in a tubular, 2-cleft spathe. Staminate flowers stalked, with a perianth of 3 petal-like lobes and 9 distinct stamens. Pistillate flowers nearly sessile in the spathe, with a 6- to 12-celled inferior ovary with 9 to 12 stigmas. Fruit stalked, with warty surface, many-seeded, opening by valves. — A few species in tropical marine waters.

Thalassia testudinum Konig. Turtle grass, "Seaweed." Figs. 50; 48a, C. Map 94.

Common in salt water in shallow bays and about reefs off the coast of southern Florida and along the Gulf Coast. Forming dense and extensive submarine meadows. During storms large windrows of it are washed on the beaches forming the "wrack" or "seaweed" of the region.

Map 94. *Thalassia testudinum.*

Map 95. *Vallisneria americana.*

VALLISNERIA

Dioecious, submersed perennials from creeping rootstocks with fibrous roots and fleshy propagating buds. Leaves basal in clusters at the

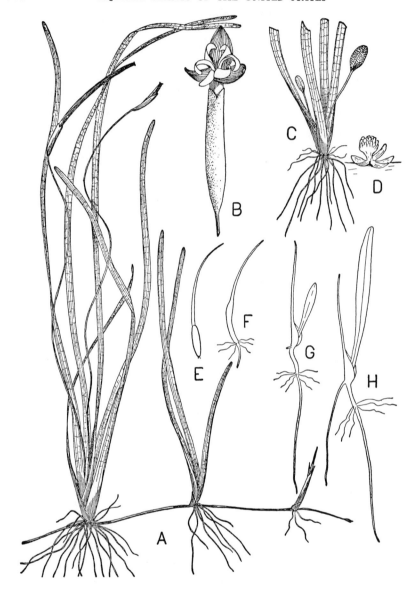

Fig. 48. *Vallisneria americana.*

(A) Pistillate plant with young fruits and rhizome; x 1/3. (B) Pistillate flower; x 2. (C) Base of staminate plant with inflorescences; x 1/2. (D) Staminate flower; x 4. (E) Germinating seed; x 3. (F, G, H) Seedling in different stages of development; x 3.

nodes, linear, ribbon-like, up to 1 meter long, the upper part often floating on the water. Staminate spadix borne on a short peduncle and inclosed in a 2- or 3-parted spathe, the flowers numerous, becoming detached and floating on the water; perianth of 3 sepals; stamens usually 3. Pistillate flowers solitary in a 2-cleft spathe borne at the apex of a long, slender scape, floating on the water; sepals 3, fused to the inferior ovary; petals 3; pistil with three 2-lobed, nearly sessile stigmas; ovary long-cylindric, 1-celled, with 3 parietal placentae. Fruit berrylike, about 5 to 10 cm. long, with numerous seeds imbedded in a gelatinous mass.

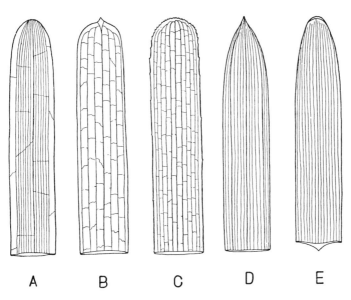

FIG. 48a. *Vallisneria americana* (A), *Zostera marina* (B), *Thalassia testudinum* (C), *Potamogeton zosteriformis* (D), *Sparganium fluctuans* (E). Diagrams showing differences in the tapelike leaves of several species.

KEY TO SPECIES OF VALLISNERIA

1. Stigmas 2-cleft for less than ½ their length; sepals 2 to 3 mm. long; leaves mostly 5 to 6 mm. wide .... *V. americana*
1. Stigmas 2-cleft to near the base; sepals 4 to 6 mm. long; leaves mostly 15 to 20 mm. wide .... *V. neotropicalis*

Vallisneria americana Michx. Wild celery, "Eelgrass." Figs. 48; 48a, A. Map 95.

Common in shallow lakes and streams throughout the eastern states.

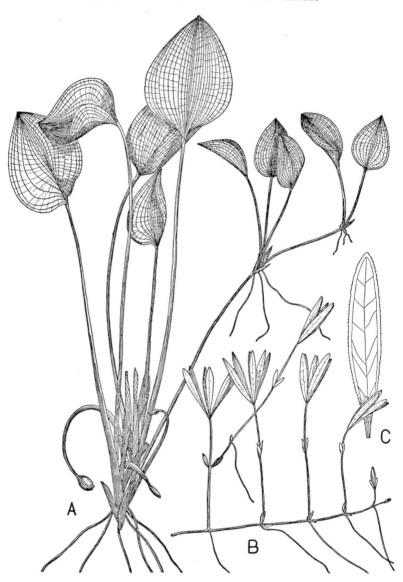

Fig. 49. *Limnobium spongia* (A) and *Halophila engelmanni* (B, C).
(A) Plant; x 1/2.
(B) Part of plant showing creeping habit; x 1/2. (C) Leaf; x 2.

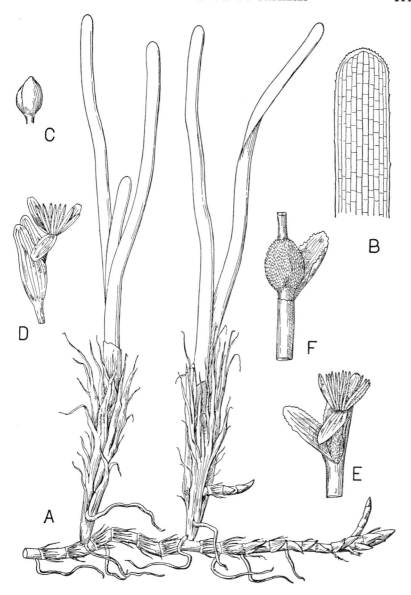

FIG. 50. *Thalassia testudinum.*

(A) Sterile plant showing habit and rhizome; x 1/2. (B) Apex of leaf; x 1 1/2. (C) Unopened staminate flower; x 1. (D) Staminate flower with 1 sepal turned down; x 1. (E) Pistillate flower with spathe cut open to show ovary with 12 stigmas; x 1. (F) Young fruit with part of spathe removed; x 1. (C-F after Rydberg, 1909.)

NOTE. The fleshy, tuber-like, thickened ends of the rhizomes and also the fleshy fruits of this species are an important source of food for waterfowl. It is often planted in lakes and ponds to improve the food supply in feeding areas.

Vallisneria neotropicalis Marie-Vict.
Local in shallow water. Florida and Cuba.

REFERENCES

Wiegand, K. M., and A. J. Eames. *Elodea*. In, Flora of Cayuga Lake Basin, New York. Cornell Univ. Mem. 92:55. 1926.
Marie-Victorin, F. L'Anacharis canadensis. Contr. Lab. Bot. Univ. Montréal 18:1-43. 1931.
———. Les Vallisnéries américaines. Contr. Inst. Bot. Univ. Montréal 46:1-38. 1943.
Wylie, R. B. A long-stalked *Elodea* flower. Univ. Iowa, Lab. Nat. Hist. Bull. 6:43-52. 1911-1913.
St. John, H. The genus *Elodea* in New England. Rhodora 22:17-29. 1920.
Wylie, R. B. The pollination of *Vallisneria spiralis*. Bot. Gaz. 63:135-145. 1917.
Fernald, M. L. The diagnostic character of *Vallisneria americana*. Rhodora 20:108-110. 1918.
Rydberg, P. A. The flowers and fruits of the turtle-grass [*Thalassia*]. N. Y. Bot. Gard. Jour. 10:261-269. 1909.

GRAMINEAE: GRASS FAMILY

Plants with fibrous roots, nodose stems usually with hollow internodes, and narrow, parallel-veined, 2-ranked leaves composed of a sheath and blade. The sheath is wrapped around the stem with its margins overlapping but usually free. At the upper end of the sheath where it is joined with the blade usually occurs a membranous or hairy fringe, the ligule.

Flowers perfect or imperfect, arranged in 2-ranked spikelets which are clustered in panicles, racemes, or spikes. Spikelets with a shortened axis (rhachilla) on which are attached a few to many bracts (the glumes), the 2 lower of which, the empty glumes, are without flowers in their axils; the succeeding ones, the lemmas or flowering glumes, each bear a solitary axillary flower and palea on a short stalk. The midrib of the glumes may project into a long bristle or awn beyond the body of the glume. In some forms the upper or lower lemmas may be reduced or sterile.

Each perfect flower consists of 3 stamens, with filaments attached apparently at the middle of the 2-celled anthers, and a solitary pistil, with 1-celled, 1-ovuled ovary and, usually, 2 styles with plumose stigmas. Fruit a grain or caryopsis, often inclosed in, or surrounded by, the glume and palea. Embryo small, along 1 side of the copious starchy endosperm. — A large family of about 7000 species in about 500 genera. Widely distributed on nearly all land surfaces of the earth; only a few species grow in water.

KEY TO GENERA

1. Fertile flower solitary in a spikelet.
   2. Spikelets with sterile flowers or glumes attached below the fertile flower.
      3. Spikelets inserted at the nodes of a thickened, articulate rhachis (forming a spikelike raceme) .... *Manisuris*
      3. Spikelets in a panicle or spike; rhachis not thickened or articulate.
         4. Spikelets with 2 sterile flowering glumes below the fertile flower .. *Phalaris*
         4. Spikelets with 1 sterile flowering glume below the fertile flower.
            5. Glumes or sterile lemma awned; spikelets in a 1-sided panicle .... *Echinochloa*
            5. Glumes or sterile lemma not awned.
               6. Spikelets attached on side of rhachis of single spike or cluster of spikes.
                  7. Spikelets dorsally compressed, lanceolate, acuminate .... *Reimarochloa*
                  7. Spikelets plano-convex, obtuse .... *Paspalum*
               6. Spikelets in an open panicle .... *Panicum*
   2. Spikelets with sterile flowers or glumes attached above the fertile flower.
      8. Spikelets unisexual.
         9. Staminate and pistillate flowers borne in the same panicle; plants robust, 1 to 4 meters tall .... *Zizania*
         9. Staminate and pistillate flowers borne in separate inflorescences; plants slender, low.
            10. Inflorescence a raceme with few flowers; plant not stoloniferous .... *Hydrochloa*
            10. Inflorescence a panicle; plant stoloniferous .... *Luziola*
      8. Spikelets perfect.
         11. Spikelets articulate below the glumes.
            12. Spikelets in an open panicle .... *Leersia*
            12. Spikelets in a dense spikelike panicle .... *Alopecurus*
         11. Spikelets articulate above the glumes.
            13. Spikelets attached in an open panicle .... *Calamagrostis*
            13. Spikelets attached in two rows in a spike.
               14. Spikelets on opposite sides of the rhachis, solitary at each node; spike solitary .... *Pholiurus*
               14. Spikelets on one side of the rhachis; spikes mostly several in a cluster.

15. Glumes equal, broad, boat-shaped .... *Beckmannia*
15. Glumes unequal, narrow .... *Spartina*
1. Fertile flowers 2 to many in a spikelet.
    16. Stems 2 to 4 meters tall, reedlike; panicle plumelike; rhachilla hairy ......
    .... *Phragmites*
    16. Stems lower, rarely 2 meters tall.
        17. Plants dioecious, creeping perennials, of salt or alkaline marshes.
            18. Inflorescence a narrow, exserted panicle; rhizome with erect branches
            .... *Distichlis*
            18. Inflorescence obscure, usually inclosed in the leaf sheath; stem creeping .... *Monanthochloë*
        17. Plants monoecious.
            19. Spikelet 2-flowered; lemma 3-nerved .... *Catabrosa*
            19. Spikelet 3- to many-flowered; lemma 5- to 9-nerved.
                20. Spikelets in racemes .... *Pleuropogon*
                20. Spikelets in panicles.
                    21. Spikelets 3- to 4-flowered; lemma hairy on callus .... *Fluminea*
                    21. Spikelets mostly 4- to many-flowered; lemma not hairy on callus ..
                    .... *Glyceria*

ALOPECURUS: Foxtail

Annuals or perennials; stems slender, erect or with decumbent base; leaf blades flat, often floating. Spikelets 1-flowered, laterally compressed, in dense, spikelike panicles. Glumes equal, fused at base, with ciliate keel, about as long as the 5-nerved, dorsally awned lemma. Awn short or several times as long as the lemma. Palea wanting.

KEY TO SPECIES OF ALOPECURUS

1. Spike bristly; awns exserted 2 to 3 mm.; stems decumbent and rooting at the lower nodes .... *A. geniculatus*
1. Spike not bristly; awns about as long as the glumes; stems usually not rooting at the lower nodes .... *A. aequalis*

Alopecurus aequalis Sobol. Short-awn foxtail. Fig. 51, A–B. Map 96.

In shallow ponds, ditches, and sloughs subject to desication. Widely distributed except in the South.

Alopecurus geniculatus L. Water foxtail. Fig. 51, C–D. Map 97.

In ponds, streams, and temporary pools. New England to Virginia, Rocky Mountains to Pacific Coast.

BECKMANNIA: Slough Grass

Annuals with stout, erect stems about 1 meter tall, terminating in a slender, interrupted panicle of short, appressed spikes. Leaves with flat

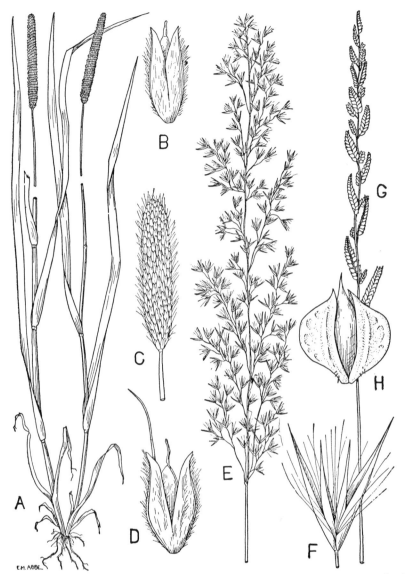

Fig. 51. *Alopecurus aequalis* (A, B), *A. geniculatus* (C, D), *Calamagrostis canadensis* (E, F), and *Beckmannia syzigachne* (G, H).

(A) Plant; x 1/2. (B) Spikelet; x 10.
(C) Inflorescence; x 1. (D) Spikelet; x 10.
(E) Panicle of spikelets; x 1. (F) Spikelet; x 10.
(G) Inflorescence, x 1/2. (H) Spikelet; x 10.

blades. Spikelets about 3 mm. long, nearly as wide, laterally compressed, nearly sessile, 1- or 2-flowered, arranged in 2 rows on 1 side of a slender rhachis. Glumes equal, inflated, keeled, 3-nerved. Lemmas 5-nerved, narrow, protruding slightly.

Beckmannia syzigachne (Steud.) Fernald. Fig. 51, G–H. Map 98.

In ditches and along streams, ponds, and marshes. Widespread in the northern states. This species has 1-flowered spikelets. The European *B. eruciformis* (L.) Host. has 2-flowered spikelets.

CALAMAGROSTIS: REED GRASS

Perennials, mostly with creeping rootstocks and erect stems bearing open or narrow panicles of 1-flowered spikelets. Leaves with flat blades. Glumes nearly equal, slender, exceeding the 5-nerved, awned lemma. Rhachilla prolonged behind the palea into a short, hairy bristle. Callus hairs abundant, often as long as the lemma. — Of the 20 or more na-

MAP 96. *Alopecurus aequalis.*

MAP 97. *Alopecurus geniculatus.*

MAP 98. *Beckmannia syzigachne.*

MAP 99. *Calamagrostis canadensis.*

tive species occurring in the United States, all are terrestrial except the following which grows in marshes and water.

Calamagrostis canadensis (Michx.) Beauv. Bluejoint grass. Fig. 51, E–F. Map 99.

Common in marshes, wet meadows, and beaver ponds, mostly on mucky or peaty soil. A variable and widespread grass throughout the northern United States, less common southward and absent in the extreme South.

CATABROSA: Brook Grass

Aquatic perennials with creeping or decumbent stems and short, flat leaf blades; glabrous throughout. Panicle erect, up to 20 cm. long, with whorls of spreading branches. Spikelets mostly 2-flowered, about 3 mm. long. Glumes unequal, flat, without nerves, coarsely toothed at the blunt apex, shorter than the broad, 3-nerved lower lemma.

Catabrosa aquatica (L.) Beauv. Map 100.

In streams, spring holes, and wet meadows, chiefly in the mountains. Rocky Mountain states and northward.

Map 100. *Catabrosa aquatica.*

Map 101. *Distichlis spicata.*

DISTICHLIS: Salt Grass

Low perennials with extensively creeping, scaly rhizomes and short, stiff, erect stems and prominently 2-ranked leaves with short blades and overlapping sheaths. Plants dioecious, with short, dense panicles of few spikelets. Spikelets several- to many-flowered. Glumes unequal, broad, acute, keeled, 3- to 7-nerved. Lemmas firm, closely overlapping, with 9 to 11 faint nerves. Palea as long as the lemmas.

Distichlis spicata (L.) Greene. Seashore salt grass. Fig. 52, A–B. Map 101.

Common on salt marshes, mostly near the seashore. Maine to Florida and Texas; Pacific Coast.

ECHINOCHLOA

Annuals with coarse, succulent stems. Leaves with flat blades and compressed sheaths without ligules. Spikelets plano-convex, crowded on 1-sided racemes arranged in large panicles. First glume about half

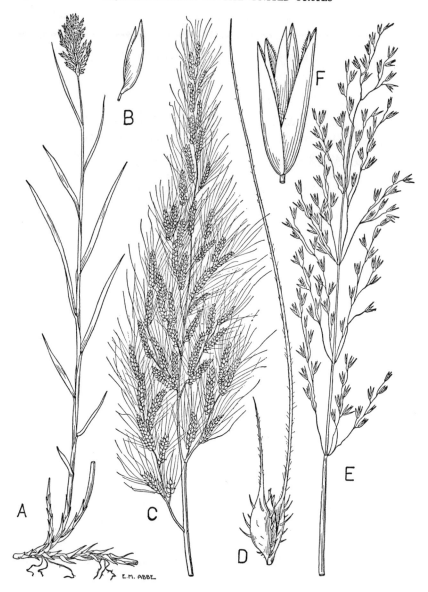

FIG. 52. *Distichlis spicata* (A, B), *Echinochloa walteri* (C, D), and *Fluminea festucacea* (E, F).

(A) Plant; x 1/2. (B) Spikelet; x 5.
(C) Panicle; x 1/2. (D) Spikelet; x 5.
(E) Panicle; x 1/2. (F) Spikelet; x 5.

as long as the spikelet; second glume and the lower lemma equal, the latter sterile (with stamens) or neutral (with neither stamens nor pistil), usually awned. Fertile lemma plano-convex, glossy.

KEY TO SPECIES OF ECHINOCHLOA

1. Sterile flower staminate .... *E. paludigena*
1. Sterile flower neutral.
    2. Lower sheaths mostly hispid or glabrous; spikelets long-awned (1 to 3 cm.), in dense panicle .... *E. walteri*
    2. Lower sheaths glabrous; spikelets short-awned (1 cm.) ...... *E. crus-pavonis*

Echinochloa crus-pavonis (HBK.) Schult. Map 102.

Marshes, sloughs, and ditches. Lowlands along the Gulf Coast; local in Virginia.

Echinochloa paludigena Wiegand. Map 103.

In shallow water in ditches and marshes. Florida.

Echinochloa walteri (Pursh) Heller. Fig. 52, C–D. Map 103.

In shallow water, sloughs, and intermittent ponds. Widespread along the Atlantic Coast and in the Mississippi Valley.

MAP 102. *Echinochloa crus-pavonis*.   MAP 103. *Echinochloa paludigena*, ▲; *E. walteri*, ●.

FLUMINEA

Perennials with tall, succulent stems and fleshy, creeping rhizomes. Leaves flat, scabrous on the upper surface, long, taper-pointed. Spikelets in open terminal panicles, 3- or 4-flowered, about 8 mm. long. Glumes nearly equal, somewhat scarious, the first 3-nerved, the second 5-nerved, about as long as the lower lemma. Lemmas firm, rounded on the back, with 7 unequal nerves, lacerate at apex. Palea narrow, flat, as long as the lemma.

Fluminea festucacea (Willd.) Hitchc. Fig. 52, E–F. Map 104.

Shallow water, streams, marshes, and sloughs. Northwestern states; also eastern Oregon.

MAP 104. *Fluminea festucacea.*

GLYCERIA: MANNA GRASS

Perennials with tall, slender, simple stems from rhizomes or from the rooted, decumbent, or creeping basal part of the stem. Leaves with broad, flat blades and closed or partially closed sheaths. Inflorescence an open or crowded panicle. Spikelets few- to many-flowered, nearly cylindrical or somewhat flattened; glumes unequal, short, often scarious and mostly 1-nerved; lemmas firm, rounded on the back and apex, with 5 to 9 parallel nerves. — Many species, mostly growing in marshy places or in shallow water.

KEY TO SPECIES OF GLYCERIA

1. Spikelets 10 mm. long or longer, cylindrical; panicle erect and narrow.
  2. Lemmas acute at apex, exceeded by the palea .....*G. acutiflora*
  2. Lemmas obtuse at apex, scarcely exceeded by the palea.
    3. Surface of lemma glabrous between the nerves ....*G. borealis*
    3. Surface of lemma hairy between the nerves.
      4. Length of lemma about 3 mm. ....*G. leptostachya*
      4. Length of lemma 4 to 7 mm.
        5. Lemma green or pale yellow-green ....*G. septentrionalis*
        5. Lemma purple, at least near the apex ....*G. occidentalis*
1. Spikelets 2 to 7 mm. long, ovate to oblong, somewhat flattened; panicle mostly nodding.
  6. Lemmas 5-nerved ....*G. pauciflora*
  6. Lemmas 7-nerved.
    7. Spikelets in crowded, narrow panicles.
      8. Panicle short, stiff, erect, rarely 10 cm. long ....*G. obtusa*
      8. Panicle slender, lax, nodding, mostly 15 to 25 cm. long .....*G. melicaria*
    7. Spikelets in open, spreading panicles.
      9. Nerves of lemma inconspicuous; spikelets more than 3 mm. wide
        ....*G. canadensis*
      9. Nerves of lemma conspicuous; spikelets less than 3 mm. wide.

10. Stems decumbent at base.
    11. Leaf blade narrow, 1 to 3 mm. wide .... G. fernaldii
    11. Leaf blade broad, 4 to 8 mm. wide .... G. pallida
10. Stems erect.
    12. Second glume 2 to 2.5 mm. long; spikelets 4 to 7 mm. long
        .... G. grandis
    12. Second glume 1 mm. long; spikelets short, 3 to 4 mm. long.
        13. Blades mostly 2 to 4 mm. wide, often folded .... G. striata
        13. Blades 6 to 12 mm. wide, flat .... G. elata

Glyceria acutiflora Torr. Fig. 53, K. Map 105.
Shallow water and marshes. Northeastern states.

Glyceria borealis (Nash) Batch. Fig. 53, C. Map 106.
Ponds, slow streams, and springy places. Throughout the northern states and southward in the Rocky Mountains.

Glyceria canadensis (Michx.) Trin. Fig. 53, A-B. Map 107.
Marshes and boggy places. Northeastern states westward to Minnesota.

Glyceria elata (Nash) Hitchc. Fig. 53, D. Map 108.
Marshes and springy places. Western states.

Glyceria fernaldii (Hitchc.) St. John. Fig. 53, E. Map 108a.
In shallow water and wet mucky places. New England and Great Lakes region.

Glyceria grandis Wats. Fig. 53, F. Map 109.
In marshes and along shores of streams and ponds. Widespread throughout the northern states.

Glyceria leptostachya Buckl. Map 110.
In shallow water and marshes. Pacific Coast states.

Glyceria melicaria (Michx.) T. F. Hubb. Fig. 53, G. Map 111.
Swamps and wet woodlands. Chiefly in the northeastern and Middle Atlantic states.

Map 105. *Glyceria acutiflora.*

Map 106. *Glyceria borealis.*

MAP 107. *Glyceria canadensis.*

MAP 108. *Glyceria elata.*

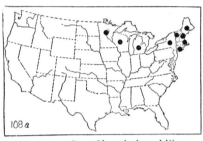

MAP 108a. *Glyceria fernaldii.*

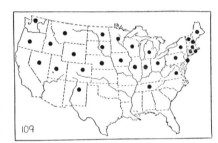

MAP 109. *Glyceria grandis.*

MAP 110. *Glyceria leptostachya.*

MAP 111. *Glyceria melicaria.*

Glyceria obtusa (Muhl.) Trin. Fig. 53, H. Map 112.

Shallow ponds, bogs, and springy places. Mostly on the Coastal Plain, from Maine to Virginia.

Glyceria occidentalis (Piper) J. C. Nelson. Fig. 53, I. Map 112.

Shallow ponds, slow streams, and marshes. Chiefly in the Pacific Northwest.

Glyceria pallida (Torr.) Trin. Fig. 53, J. Map 113.

In cold bogs, ponds, and slow streams. Chiefly in the northeastern states, westward to Minnesota and Missouri.

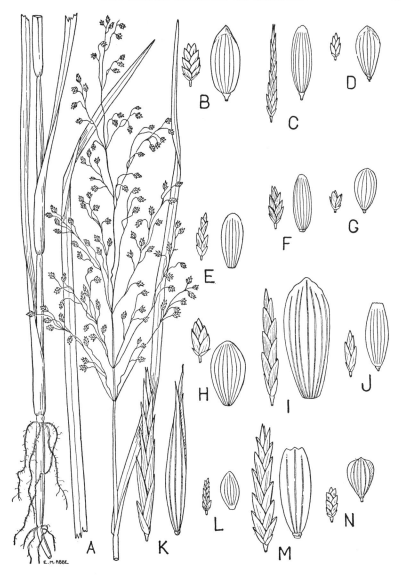

FIG. 53. *Glyceria canadensis* (A, B), *G. borealis* (C), *G. elata* (D), *G. fernaldii* (E), *G. grandis* (F), *G. melicaria* (G), *G. obtusa* (H), *G. occidentalis* (I), *G. pallida* (J), *G. acutiflora* (K), *G. pauciflora* (L), *G. septentrionalis* (M), and *G. striata* (N). Plant and inflorescence, x 1/2. Spikelets, x 2. Glumes, x 6.

MAP 112. *Glyceria occidentalis*, West; *G. obtusa*, East.

MAP 113. *Glyceria pallida*.

MAP 114. *Glyceria pauciflora*.

MAP 115. *Glyceria septentrionalis*.

MAP 115a. *Glyceria striata*.

MAP 116. *Hydrochloa carolinensis*.

Glyceria pauciflora Presl. Fig. 53, L. Map 114.

In shallow water and marshes. Mostly from the Rocky Mountains westward.

Glyceria septentrionalis Hitchc. Fig. 53, M. Map 115.

Shallow ponds and marshes. In the eastern states, westward to the Mississippi Valley.

Glyceria striata (Lam.) Hitchc. Fig. 53, N. Map 115a.

In bogs, wet meadows, and shores of streams and ponds. Widespread throughout the United States.

HYDROCHLOA

Aquatics with branched stems up to 1 meter long, decumbent or creeping at the base and the upper part often floating. Leaves with flat blades often floating. Spikelets without glumes, 1-flowered, unisexual, only a few in a cluster; the staminate in small, terminal racemes and the pistillate in short, axillary racemes. Staminate spikelet with a 7-nerved lemma and a 2-nerved palea inclosing 6 stamens. Pistillate spikelet with a 7-nerved lemma and 5-nerved palea, bearing a pistil with slender stigmas.

Hydrochloa carolinensis Beauv. Fig. 54, D. Map 116.

In ponds, pools, and sluggish streams. On the lowlands along the South Atlantic and Gulf Coast.

LEERSIA: Cut-Grass

Perennials with slender, creeping rhizomes and weak, usually decumbent stems rooting at the lower nodes. Leaf blades flat, coarsely scabrous, especially along the margins. Spikelets 1-flowered, without glumes. Lemma compressed laterally, boat-shaped, with a prominent keel, mostly 5-nerved, the keel and 2 lateral nerves mostly hispid or ciliate; palea mostly 3-nerved, as long as the lemma, with hispid-ciliate keel. Stamens 6 or fewer.

KEY TO SPECIES OF LEERSIA

1. Branches of panicle appressed or ascending, panicle narrow .... *L. hexandra*
1. Branches of panicle spreading when mature, panicle open .... *L. oryzoides*

Leersia hexandra Swartz. Map 117.

Shallow water of ponds, ditches, and sloughs. Mostly near the coast from North Carolina to Florida and Texas.

Map 117. *Leersia hexandra*.

Map 118. *Leersia oryzoides*.

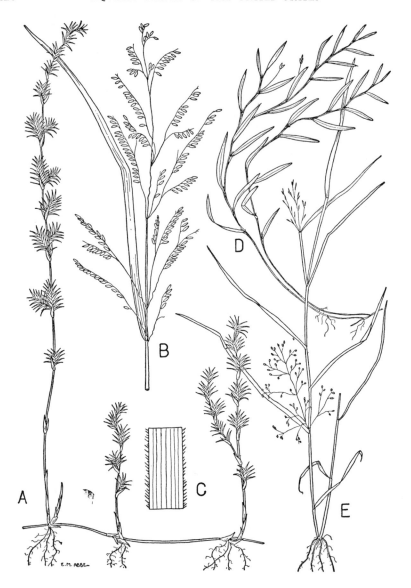

FIG. 54. *Monanthochloë littoralis* (A), *Leersia oryzoides* (B, C), *Hydrochloa carolinensis* (D), and *Luziola bahiensis* (E).

(A) Plant with rhizome; x 1/2.
(B) Inflorescence; x 1/2. (C) Section of blade with spiny-toothed margin; x 2.
(D) Part of a small plant; x 1/2.
(E) Plant; x 1/2.

Leersia oryzoides (L.) Swartz. Fig. 54, B–C. Map 118.

Common in shallow ponds, intermittent pools, and wet places bordering streams. Widespread east of the Great Plains; infrequent westward.

LUZIOLA

Low perennials with creeping or slender erect stems. Spikelets without glumes, 1-flowered, unisexual, the staminate and pistillate spikelets borne in separate, terminal, or also axillary, panicles on the same plant. Lemma about as long as the palea, with several to many nerves. Staminate flower with 6 or, rarely, more stamens; pistillate flower forming a globose grain free from the lemma.

KEY TO SPECIES OF LUZIOLA

1. Pistillate spikelets 2 mm. long, on the same shoot as the staminate spikelets .... L. peruviana
1. Pistillate spikelets 4 to 5 mm. long, on separate shoots from the staminate spikelets .... L. bahiensis

Luziola bahiensis (Steud.) Hitchc. Fig. 54, E. Map 119.

Local in lagoons and along banks of streams near the coast in Alabama and southward.

Luziola peruviana Gmel. Map 120.

Local in wet depressions and on muddy shores along the Gulf Coast and southward.

Map 119. *Luziola bahiensis.*   Map 120. *Luziola peruviana.*

MANISURIS

Perennials with tall, erect, compressed, freely branching stems from long, creeping rootstocks. Spikelets in pairs at the crowded nodes of a thickened, jointed rhachis, forming a flattened raceme. Each pair of spikelets containing 1 sessile and perfect and 1 stalked and sterile or rudimentary spikelet. First glume coriaceous, with keels winged near the apex; lemma and palea of fertile spikelet thin and hyaline.

Manisuris altissima (Poiret) Hitchc. Map 121.

In ponds, slow streams, and ditches in warm temperate and tropical regions. Introduced in America. From southern Texas southward.

MONANTHOCHLOË

Dioecious perennials with tough, slender, creeping or prostrate stems with short, erect branches bearing clusters of leaves with stiff, awl-like blades about 1 cm. long. Spikelets solitary or a few in the leaf axils, 3- to 5-flowered, without glumes, the upper flowers mostly rudimentary. Lemmas rounded on the back, convolute about the pistil, several-nerved; palea 2-nerved.

Map 121. *Manisuris altissima.*   Map 122. *Monanthochloë littoralis.*

Monanthochloë littoralis Engelm. Salt-flat grass. Fig. 54, A. Map 122

On muddy and sandy shores and on tidal flats along seashores. From the southern parts of Florida, Texas, and California to Central America.

PANICUM

Annuals or perennials with panicles or, rarely, racemes of spikelets. Spikelets more or less compressed dorsiventrally, with 1 perfect terminal flower and a sterile or staminate flower below. Glumes 2, herbaceous, mostly quite unequal, the lower minute, the second as long as the sterile lemma which bears a membranous or hyaline palea and sometimes also a staminate flower in its axil. Lemma bearing the fertile flower hard, obtuse, nerveless, with margins inrolled over the hard palea. — A large genus of many terrestrial species and a few aquatics. The following perennial species have spikelets in 1-sided, spikelike racemes.

KEY TO SPECIES OF PANICUM

1. Stems in tufts; spikelets 2 to 25 mm. long .... P. geminatum
1. Stems from elongate rhizomes; spikelets about 3 mm. long .... P. paludivagum

Panicum geminatum Forsk. Fig. 55, C. Map 123.

In shallow water and on wet lowlands mostly near the coast. Southern Florida, Louisiana, and Texas southward.

MAP 123. *Panicum geminatum.*   MAP 124. *Panicum paludivagum.*

Panicum paludivagum Hitchc. Fig. 55, A–B. Map 124.

In fresh water of lakes and streams. Florida to Texas and southward into Central America.

PASPALUM

Mostly perennials with 1 or more 1-sided, spikelike racemes. Racemes solitary or 2 or more on a common axis. Spikelets solitary or in pairs, in 2 rows on 1 side of the rhachis, obtuse, plano-convex, with a terminal fertile lemma and a basal sterile lemma; glumes both wanting or the lower wanting and the second much shorter than the fertile lemma. — Mostly terrestrial species but the following grow in water.

KEY TO SPECIES OF PASPALUM

1. Racemes falling entire from the axis; stems 1 to 2 meters long, creeping or floating; leaf sheaths inflated .... P. repens
1. Racemes persisting on the axis; stems up to 1 meter long, often creeping; leaf sheaths scarcely inflated.
    2. Spikelets 2 mm. long, ovate to obovate .... P. dissectum
    2. Spikelets at least 3 mm. long, pointed .... P. acuminatum

Paspalum acuminatum Raddi. Map 125.

In shallow ponds and streams, and in marshy places. From southern Louisiana and Texas southward.

130 AQUATIC PLANTS OF THE UNITED STATES

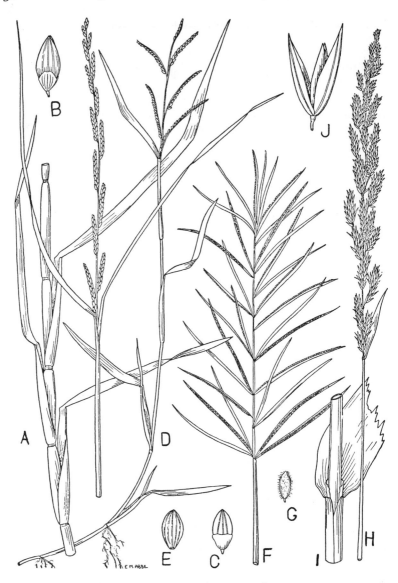

FIG. 55. *Panicum paludivagum* (A, B), *P. geminatum* (C), *Paspalum dissectum* (D, E), *P. repens* (F, G), and *Phalaris arundinacea* (H–J).

(A) Plant; x 1/2. (B, C) Spikelets; x 5.
(D) Plant; x 1/2. (E) Spikelet; x 5.
(F) Inflorescence; x 1/2. (G) Spikelet; x 5.
(H) Inflorescence; x 1/2. (I) Leaf sheath and ligule; x 2. (J) Spikelet; x 5.

MAP 125. *Paspalum acuminatum.*

MAP 126. *Paspalum dissectum.*

Paspalum dissectum L. Fig. 55, D–E. Map 126.
On muddy or sandy shores of ponds and streams. Along the South Atlantic Coast and in the lower Mississippi Valley.

Paspalum repens Berg. Fig. 55, F–G. Map 127.
In shallow ponds and sluggish streams; sometimes forming floating mats. Chiefly in the southeastern states.

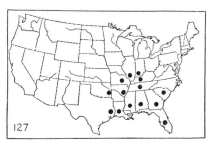
MAP 127. *Paspalum repens.*

PHALARIS: Canary Grass

Perennials or annuals with erect stems, flat leaf blades, and narrow or spikelike panicles. Spikelets laterally compressed, with a terminal, coriaceous lemma and a perfect flower, and usually with 2 reduced, scalelike, sterile lemmas below. The 2 glumes equal, wing-keeled, longer than the fertile lemma. — Mostly terrestrial species but the following perennial species with long rhizomes may grow in areas subject to inundation.

Phalaris arundinacea L. Reed canary grass. Fig. 55, H–J. Map 128.
In marshes and wet meadows and along streams and ditches. Widespread except in the southeastern states. In many places it has been introduced as a forage grass on marshy meadows.

MAP 128. *Phalaris arundinacea*.

MAP 129. *Pholiurus incurvus*.

PHOLIURUS

Low annuals with tufted, spreading or ascending stems with curved, cylindric spikes. Spikelets 1- or 2-flowered, imbedded in the jointed, cylindric rhachis of the spike, the mature rhachis breaking into segments each with an attached spikelet. Glumes 2, on the front of the spikelet, coriaceous, 5-nerved, acute; lemma with its back to the rhachis, shorter than the glumes, 1-nerved.

Pholiurus incurvus (L.) Schinz. and Thell. Sickle-grass. Map 129.

Local in salt marshes and tidal flats. Middle Atlantic Coast and also in California and Oregon. Introduced from Europe.

PHRAGMITES: REED

Coarse perennials with hard, erect stems 1 to 5 meters high, from long, coarse, scaly, creeping rootstocks. Leaves with long, flat blades. Spikelets in plumelike, terminal panicles 15 to 40 cm. long. Spikelets several-flowered, the lower flower staminate or neutral. Glumes 3-nerved or the second 5-nerved, the first about ½ as long as the second. Lemmas 3-nerved, acuminate, glabrous, the lowest longer than the second glume, those of the upper part of the spikelet successively shorter. Palea much shorter than the lemma. The rhachilla with silky hairs which finally exceed the glumes and lemmas.

Phragmites communis Trin. Reed. Fig. 56, A–B. Map 130.

Along shores of lakes and streams, in fresh-water and brackish marshes, and about springs. Widespread except in the southeastern states.

PLEUROPOGON: SEMAPHORE GRASS

Annuals or perennials with simple stems bearing large spikelets in a loose raceme with slender, flexuous axis. Spikelets several-flowered,

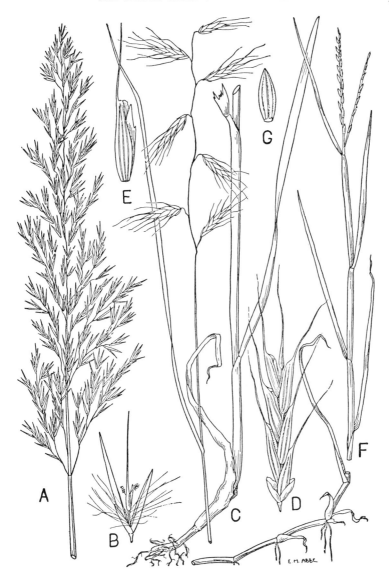

FIG. 56. *Phragmites communis* (A, B), *Pleuropogon refractus* (C–E), and *Reimarochloa oligostachya* (F, G).

(A) Panicle; x 1/2. (B) Spikelet; x 3.
(C) Part of plant with spikelets; x 1/2. (D) Spikelet; x 2. (E) Glume with palea; x 3.
(F) Part of plant; x 1/2. (G) Spikelet; x 3.

linear; glumes unequal, scarious or lacerate at the apex, the first 1-nerved, the second obscurely 3-nerved. Lemmas 7-nerved, entire or 2-toothed at apex, often ending in an awn.

KEY TO SPECIES OF PLEUROPOGON

1. Plants annual; spikelets ascending .... P. *californicus*
1. Plants perennial; spikelets becoming reflexed or drooping .... P. *refractus*

Pleuropogon californicus (Nees) Benth.
Marshes, bogs, and wet meadows. Northern to central California.

Pleuropogon refractus (Gray) Benth. Fig. 56, C-E. Map 131.
In mountain bogs, wet meadows, and streams. From the Cascade Mountains in Washington to northern California.

MAP 130. *Phragmites communis*.

MAP 131. *Pleuropogon refractus*, ●; *Reimarochloa oligostachya*, ▲.

REIMAROCHLOA

Perennials, with spreading stems rooting at the nodes. Spikelets in racemes, nearly sessile and alternate, in 2 rows along 1 side of a narrow, flattened rhachis. Spikelets with 1 fertile terminal flower and a sterile or staminate flower below; glumes wanting.

Reimarochloa oligostachya (Munro) Hitchc. Fig. 56, F-G. Map 131.
In shallow water and marshes. In southern Florida.

SPARTINA: CORD GRASS

Perennials, mostly with erect stems from stout, cordlike, scaly rhizomes. Leaves mostly long, tough, and involute at least toward the apex. Spikelets sessile, crowded on 1 side of a continuous rhachis forming a spike. Spikes few or many in a racemose cluster. Spikelets 1-flowered, laterally flattened; the glumes 1-nerved, keeled, sometimes short-awned, the outer shorter than the second; lemma shorter than the second glume, keeled; palea 2-nerved, keeled.

KEY TO SPECIES OF SPARTINA

1. Stems slender, less than 1 meter high; leaf blades slender, less than 3 mm. wide, involute .... *S. patens*
1. Stems stout, more than 1 meter high; leaf blades mostly at least 1 cm. wide, flat at least in the lower part.
   2. Spikes of inflorescence distant and spreading; second glume with a slender awn 5 to 7 mm. long .... *S. pectinata*
   2. Spikes of inflorescence close, mostly appressed; second glume without slender awn.
      3. Margin of leaf blade coarsely scabrous; glumes with scabrous keels .... *S. cynosuroides*
      3. Margin of leaf blade glabrous or minutely scabrous; glumes glabrous or ciliate on the keels.
         4. Inflorescence dense, spikelike, composed of overlapping spikes; spikelets mostly curved .... *S. leiantha*
         4. Inflorescence slender; spikelets usually straight .... *S. alterniflora*

Spartina alterniflora Lois. Fig. 57, D–E. Map 132.

In shallow water and salt marshes along the seacoast. From Maine to Texas.

Spartina cynosuroides (L.) Roth. Fig. 58, C–D. Map 133.

In salt or brackish marshes. Mostly near the seacoast from Massachusetts to Texas.

Spartina leiantha Benth. Fig. 58, E–F.

In salt marshes. Coast of California.

Spartina patens (Ait.) Muhl. Fig. 57, A–C. Map 134.

In salt marshes along the Atlantic Coast; rare in inland marshes in New York and Michigan. Extensive salt meadows of this species along the Atlantic Coast are sometimes cut to obtain the salt hay formerly used extensively for packing and bedding material.

MAP 132. *Spartina alterniflora.*

MAP 133. *Spartina cynosuroides.*

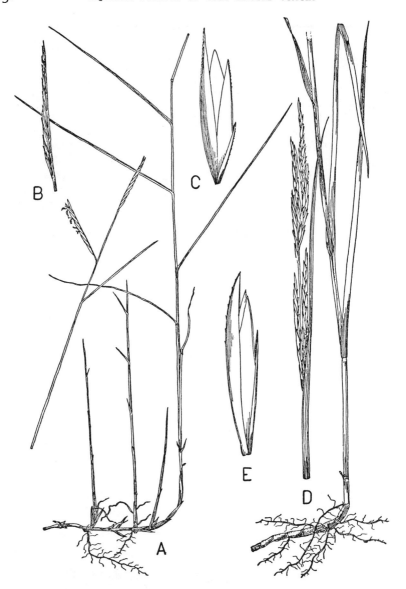

FIG. 57. *Spartina patens* (A–C) and *S. alterniflora* (D, E).

(A) Plant showing rhizome and inflorescence; x 1/2. (B) Cluster of spikelets; x 1/2. (C) Spikelet; x 5.

(D) Plant; x 1/2. (E) Spikelet; x 5.

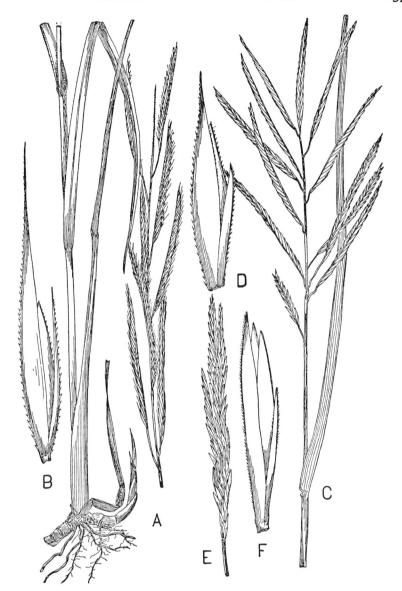

Fig. 58. *Spartina pectinata* (A, B), *S. cynosuroides* (C, D), and *S. leiantha* (E, F).

(A) Plant; x 1/2. (B) Spikelet; x 5.
(C) Inflorescence; x 1/2. (D) Spikelet; x 5.
(E) Inflorescence; x 1/2. (F) Spikelet; x 5.

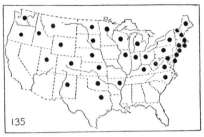

Map 134. *Spartina patens*.   Map 135. *Spartina pectinata*.

Spartina pectinata Link. Fig. 58, A–B. Map 135.

In shallow, sandy lakes and fresh-water marshes, also in brackish marshes along the Atlantic Coast. Widespread except in the South. This is an important sand binder along sandy shores of lakes and on shoals in shallow water.

ZIZANIA: Wild Rice

Annuals (or rarely perennials) with stout, erect stem and many prop roots from the lower nodes, large flat leaves, and a large terminal panicle of spikelets. Spikelets unisexual, 1-flowered, the staminate pendulous on the more or less spreading lower branches and the pistillate appressed on the upper, nearly erect branches. Spikelets with obsolete glumes reduced to a collar. Pistillate spikelets terete, or finally angular, with firm, thin, slender, 3-nerved lemma tapering into a long awn; palea 2-nerved; grain cylindric, 1 to 2 cm. long. Staminate spikelet soft; lemma 5-nerved, linear, pointed; palea 2-nerved; stamens 6.

KEY TO SPECIES OF ZIZANIA

1. Plants annual, with erect stems .... *Z. aquatica*
1. Plants perennial, with decumbent rootstocks .... *Z. texana*

Zizania aquatica L. Annual wild rice. Fig. 59. Map 136.

In shallow water of lakes, slow streams, and borders of marshes, always on soft or muddy bottom. Locally abundant in lowlands from the Atlantic Coast westward through the Mississippi Valley. The smaller plants from the northern part of its range have been separated as *Z. aquatica* var. *angustifolia* Hitchc.

NOTE. This species has been introduced in many localities to improve the food supply for waterfowl. Many attempts to establish wild rice have failed because the bottom was too hard and not fertile. Wild-rice

THE PLANTS ARRANGED BY FAMILIES 139

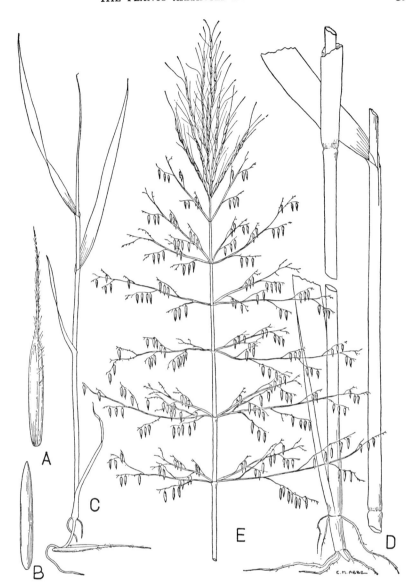

Fig. 59. *Zizania aquatica*.

(A) Grain with glume attached; x 2. (B) Grain; x 2. (C) Young seedling; x 1. (D) Lower part of plant; x 1/4. (E) Panicle with staminate flowers on lower branches and pistillate flowers on upper branches; x 1/4.

seed does not withstand drying; if air-dried the seed looses its viability. Seed intended for germination must be stored in water or in wet moss at a temperature near 1° to 3° C. until time to plant. The seed of wild rice was an important food item of the North American Indians. In the marshes about the Great Lakes region it is still gathered by some of the Indians for their own use and also to supply certain markets.

Map 136. *Zizania aquatica*.   Map 137. *Zizania texana*.

Zizania texana Hitchc. Texas wild rice. Map 137.

In flowing water in ditches and streams. San Marcos, Texas.

REFERENCES

Hitchcock, A. S. Manual of grasses of the United States. U. S. Dept. Agr., Misc. Publ. 200. 1040 pp. 1935.

Chambliss, C. E. The botany and history of *Zizania aquatica* L. ("Wild rice"). Jour. Wash. Acad. Sci. 30:185–205. 1940.

Fernald, M. L. *Phragmites communis* versus *P. maximus*. Rhodora 43:286–287. 1941.

CYPERACEAE: Sedge Family

Plants with the general appearance of grasses or rushes, with fibrous roots and mostly solid stems, often from creeping rootstocks. Leaves mostly linear, parallel-veined, basal or alternate on the stem, 3-ranked, with closed sheath. Flowers in the axils of scales (glumes) in spikelets, reduced, naked or subtended by bristles or scales, perfect or imperfect. Stamens 3, with anthers attached basally. Ovary 1-celled, 1-ovuled, with 2- or 3-cleft style. Fruit an achene, flattened, lenticular or 3-angular. Seed with small embryo and large, starchy endosperm. — About 3000 species, many of them widely distributed in wet places or in water.

KEY TO GENERA

1. Flowers all imperfect; achene inclosed in a sac (perigynium) .... *Carex*
1. Flowers perfect (sometimes the stamens or pistil aborted in some flowers).
    2. Spikelets mostly several- to many-flowered, with 1 empty lower scale.
        3. Scales of spikelets in 2 ranks.
            4. Stem scapose; spikelets in a terminal cluster; perianth bristles wanting .... *Cyperus*
            4. Stem leafy; spikelets axillary; perianth bristles present .... *Dulichium*
        3. Scales of spikelets in more than 2 ranks.
            5. Achene crowned by a tubercle (enlarged, persistent style base); spikelet solitary, terminal .... *Eleocharis*
            5. Achene without tubercle; spikelets usually several in a terminal or lateral cluster.
                6. Perianth bristles many, long exserted in fruit .... *Eriophorum*
                6. Perianth bristles 0 to 8, not long exserted in fruit .... *Scirpus*
    2. Spikelets 1- to 2-flowered, with 2 or more of the lower scales empty.
        7. Achene with a tubercle or beak, surrounded by bristles .. *Rynchospora*
        7. Achene without tubercle; bristles wanting .... *Cladium*

CAREX: Sedge

Grasslike perennials, mostly with 3-sided stems and 3-ranked leaves. Leaves often finely serrate-spiny on the margin and lower midrib. Flowers all imperfect, in spikes of many several-ranked scales; spikes often clustered in heads. Staminate and pistillate flowers borne in the same spike or in separate spikes on the same plant or on different plants. Staminate flowers with 3 or rarely 2 stamens; pistillate flowers with a solitary pistil with 2-, 3-, or 4-cleft style. Achene hard, lenticular or triangular, inclosed in a sac, the perigynium. The perigynium variable in size, shape, and length of beak. — About 1000 species widely distributed, chiefly in temperate regions. Many occur in wet meadows and swales, a few in water.

KEY TO SPECIES OF CAREX

1. Achene lenticular or plano-convex; stigmas 2.
    2. Lateral spikes sessile .... *C. aquatilis*
    2. Lateral spikes or some of them peduncled.
        3. Scales of pistillate spikes little or not at all exceeding the perigynia ...... *C. nudata*
        3. Scales of pistillate spikes at least 2 times as long as the perigynia.
            4. Perigynia glossy, not papillate .... *C. obnupta*
            4. Perigynia dull, papillate .... *C. lyngbyei*

1. Achenes triangular; stigmas 3 or rarely 4.
   5. Spikes solitary, terminal .... *C. leptalea*
   5. Spikes 2 or more on a peduncle.
      6. Perigynia hairy or scabrous, close around achene, 2-toothed.
         7. Leaf blades flat, 2 to 5 mm. wide; stem sharply 3-angled ... *C. lanuginosa*
         7. Leaf blades rounded, 1 to 2 mm. wide; stem with rounded angles ....
            .... *C. lasiocarpa*
      6. Perigynia glabrous.
         8. Style jointed to achene, not persistent.
            9. Plant with long stolons, not cespitose .... *C. limosa*
            9. Plant without stolons, cespitose .... *C. mertensii*
         8. Style continuous with achene persisting as a beak; perigynia thin and papery, more or less inflated.
            10. Staminate spike solitary .... *C. comosa*
            10. Staminate spikes 2 or more.
               11. Stem thick, spongy at base; leaves nodulose; mature perigynia horizontal or reflexed .... *C. inflata*
               11. Stem scarcely spongy at base; mature perigynia appressed or ascending .... *C. vesicaria*

Carex aquatilis Wahl. Fig. 61, D–E. Map 138.

Common in shallow ponds, sloughs, and marshes. Widespread across the northern states.

Carex comosa Boott. Fig. 61, L–M. Map 139.

In marshes, swales, and shallow water. Widespread; most common in the northeastern states.

Carex lanuginosa Michx. Fig. 60, G–H. Map 140.

In shallow water, on borders of ponds, and in marshes. Widely distributed, chiefly in the northern states.

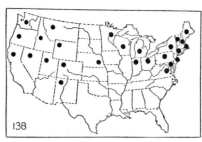

Map 138. *Carex aquatilis.*

Map 139. *Carex comosa.*

Carex lasiocarpa Ehrh. Fig. 60, A–C. Map 141.

In shallow lakes, ponds, and adjoining marshes, often making a dense turf over extensive areas. Northern states.

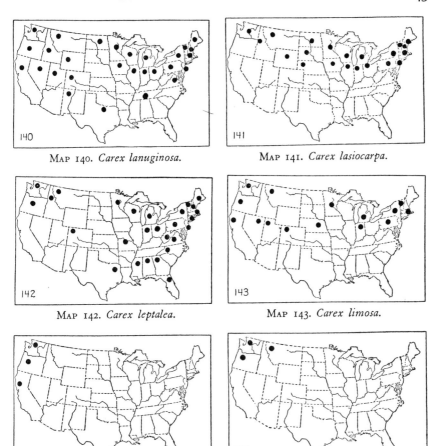

Map 140. *Carex lanuginosa*.
Map 141. *Carex lasiocarpa*.
Map 142. *Carex leptalea*.
Map 143. *Carex limosa*.
Map 144. *Carex lyngbyei*.
Map 145. *Carex mertensii*.

Carex leptalea Wahl. Fig. 61, J–K. Map 142.

Wet meadows, swales, and pools. Common in the eastern and Great Lakes states, southward in the mountains; Pacific Northwest.

Carex limosa L. Fig. 60, D–F. Map 143.

In sphagnum bogs and shallow water. Widespread across the northern United States.

Carex lyngbyei Hornem. Fig. 61, A–C. Map 144.

In brackish ponds, salt marshes, and sloughs. Pacific Coast.

Carex mertensii Prescott. Fig. 60, I–J. Map 145.

In streams and wet, boggy places, chiefly in the mountains. Pacific Northwest.

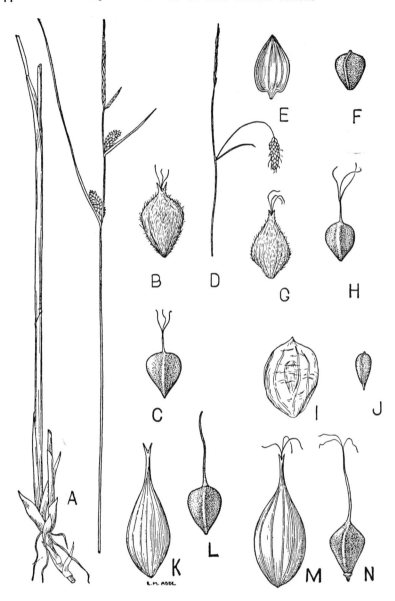

Fig. 60. *Carex lasiocarpa* (A–C), *C. limosa* (D–F), *C. lanuginosa* (G, H), *C. mertensii* (I, J), *C. inflata* (K, L), and *C. vesicaria* (M, N).

(A) Part of fruiting plant; x 1/2. (D) Staminate and pistillate spikes; x 1/2. (B, E, G, I, K, M) Perigynia; x 10. (C, F, H, J, L, N) Achenes; x 10.

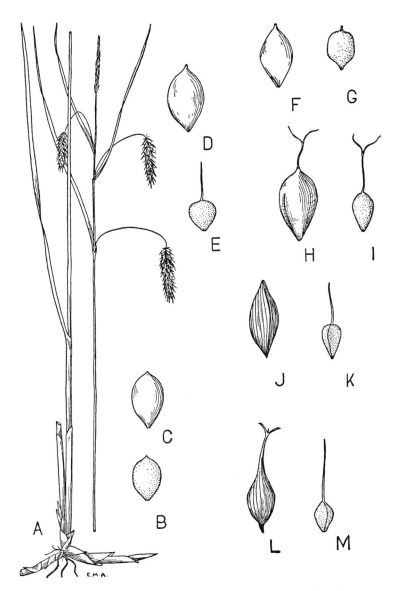

Fig. 61. *Carex lyngbyei* (A–C), *C. aquatilis* (D, E), *C. nudata* (F, G), *C. obnupta* (H, I), *C. leptalea* (J, K), and *C. comosa* (L, M)

(A) Habit of plant; x 1/2. (C, D, F, H, J, L) Perigynia; x 5. (B, E, G, I, K, M) Achenes; x 5.

Carex nudata Boott. Fig. 61, F–G. Map 146.

Among boulders and along banks of mountain streams. Western Washington to California.

Carex obnupta Bailey. Fig. 61, H–I. Map 147.

In marshes, swales, and intermittent pools. Pacific Coast states.

Map 146. *Carex nudata.*

Map 147. *Carex obnupta.*

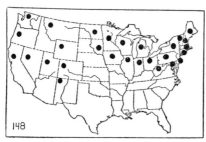
Map 148. *Carex inflata.*

Map 149. *Carex vesicaria.*

Carex inflata Huds. (*C. rostrata* Stokes). Fig. 60, K–L. Map 148.

In shallow water and marshes. Widespread, especially in the northern states. Except in the extreme north this species is represented mostly by var. *utriculosa* (Boott.) Bruce, with scales of the pistillate spikes longer and narrower than in the typical form.

Carex vesicaria L. Fig. 60, M–N. Map 149.

In shallow ponds and streams, sloughs, and marshes. Widely distributed.

CLADIUM (*MARISCUS*): Twig Rush

Perennials with creeping rhizomes and erect or decumbent, leafy or naked stems. Spikelets in cymose clusters. Spikelets oblong or ovoid, with several imbricated scales, the lower empty, the 2 or 3 upper with flowers, only the terminal flower perfect. Stamens 2; ovary with 2- to

3-cleft style. Achene ovoid or globular, somewhat corky on top, without tubercle and not subtended by bristles. — About 30 species, mostly in warm regions.

KEY TO SPECIES OF CLADIUM

1. Plant 0.3 to 1 meter high; leaves narrow, 1 to 3 mm. wide, not serrate-spiny; cymes 0.5 to 3 dm. long .... *C. mariscoides*
1. Plant 1 to 3 meters high; leaves broad, 5 to 10 mm. wide, with serrate-spiny margin and lower midrib; cluster of spikelets 3 to 9 dm. long .... *C. jamaicensis*

Cladium jamaicensis Crantz. Saw grass. Fig. 69, D–G. Map 170.

In shallow ponds, pools, wet meadows, and marshes, in fresh or brackish waters. This is the dominant plant in large areas of the Everglades of Florida. Chiefly on the Coastal Plain from Virginia to Florida and along the Gulf Coast to Texas.

Cladium mariscoides (Muhl.) Torr. Fig. 69, A–C. Map 169.

Local in shallow ponds and on sandy or boggy shores in fresh or brackish water. Northeastern states and the Great Lakes region.

CYPERUS: Galingale

Stems mostly erect, simple, triangular, leafy at base and with 1 or more leaves at the top forming an involucre to the terminal umbel or head of spikelets. Peduncles of the umbel unequal, simple or branched, sheathed at base. Spikelets few- or many-flowered, slender, mostly flattened, their scales 2-ranked, often deciduous. Flowers perfect, without bristles; stamens 1 to 3; pistil with 2- to 3-cleft, deciduous style. Achenes triangular or lenticular, without beak. — About 500 species, most abundant in warmer regions; only a few species in water.

KEY TO SPECIES OF CYPERUS

1. Involucral bracts shorter than the umbel; scapes transversely septate; rhachis winged .... *C. articulatus*
1. Involucral bracts longer than the umbel; scapes not transversely septate; rhachis not winged .... *C. ochraceus*

Cyperus articulatus L. Fig. 62, A–B. Map 150.

In shallow water and swamps in sandy regions. Southeastern and Gulf states.

MAP 150. *Cyperus articulatus.*   MAP 151. *Cyperus ochraceus.*

Cyperus ochraceus Vahl. Fig. 62, C. Map 151.

In shallow ponds and wet depressions, mostly in sandy soil. Along the Gulf Coast from Alabama to Texas.

DULICHIUM

Stems simple, jointed, hollow, terete, from perennial, creeping rootstocks. Leaves short, flat, linear, in 3 ranks, attached at the nodes all along the stem. Peduncles axillary from the leafy sheaths, with 2 ranks of sessile spikelets. Spikelets linear, flattened, with several lanceolate, decurrent scales. Flowers perfect, with 6 to 9 bristles with retrorse barbs; stamens 3; style 2-cleft. Achene flattened, linear-oblong, beaked, with persistent style. — Only 1 species.

Dulichium arundinaceum (L.) Britt. Fig. 62, D. Map 152.

In shallow water and along muddy or boggy borders of lakes, ponds, and streams; also in floating or quaking acid bogs. Widespread, chiefly in the eastern and northern states.

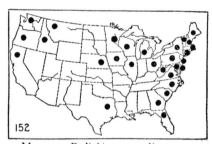

MAP 152. *Dulichium arundinaceum.*

ELEOCHARIS: Spike Rush

Perennials, or rarely annuals, with tufts of mostly leafless, erect stems (culms) from slender, creeping or matted rootstocks. Leaves basal,

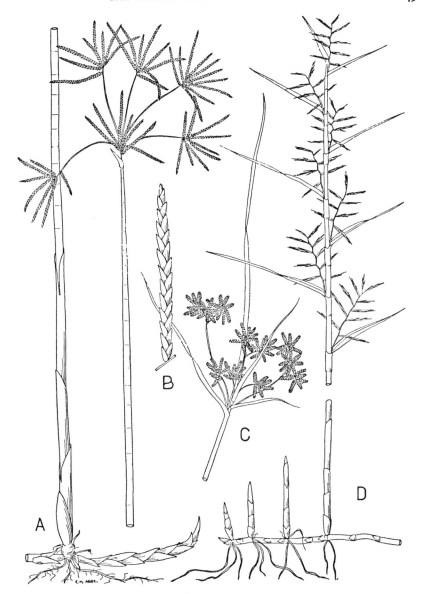

FIG. 62. *Cyperus articulatus* (A, B), *C. ochraceus* (C), and *Dulichium arundinaceum* (D).
(A) Plant; x 1/2. (B) Spikelet; x 2.
(C) Cluster of spikelets; x 1/2.
(D) Plant; x 1/2.

reduced to sheaths, or sometimes capillary. Spikelets solitary and terminal on the culms, few- to many-flowered, mostly with spirally arranged, imbricated scales. Flowers subtended by 3 to 12 (usually 6) bristles or sometimes naked; stamens 2 or 3; style 2- or 3-cleft, with bulbous base persisting as a tubercle on the triangular or lenticular achene. — About 150 species, widely distributed; mostly in marshes and shallow water.

KEY TO SPECIES OF ELEOCHARIS

1. Stems swollen, mostly as thick as the spikelets; scales firm, without prominent keel, persistent; style 2- or 3-cleft, achenes mostly lenticular.
   2. Stems septate .... *E. equisetoides*
   2. Stems nonseptate.
      3. Perianth bristles not toothed; stems terete .... *E. cellulosa*
      3. Perianth bristles toothed; stems angled.
         4. Culms stout, 2 to 5 mm. in diameter, erect, mostly 4-angled ....
            .... *E. quadrangulata*
         4. Culms slender, 1 to 2 mm. in diameter, often floating, 3-angled or flattened.
            5. Achenes about 1 mm. long .... *E. elongata*
            5. Achenes about 4 mm. long .... *E. robbinsii*
1. Stems slender, not as wide as the spikelets or, if so, then the scales of spikelets deciduous.
   6. Style 2-cleft; achenes lenticular.
      7. Plants annual, cespitose; style base compressed; achenes glossy.
         8. Tubercles nearly as wide as achene.
            9. Tubercles compressed, with concave sides .... *E. obtusa*
            9. Tubercles not more than ¼ as high as achene .... *E. engelmanni*
         8. Tubercles less than ⅔ as wide as achene.
            10. Scales pale brown; achenes black or purple-brown ...... *E. caribaea*
            10. Scales purplish-brown, with pale midvein and white margin; achenes light brown, glossy .... *E. ovata*
      7. Plants perennial, from creeping rhizomes.
         11. Stems mostly decumbent, flattened or grooved, mostly less than 2 dm. long .... *E. olivacea*
         11. Stems erect, usually stiff, terete, mostly more than 5 dm. high.
            12. Spikelets with 2 or 3 basal empty scales.
               13. Tubercle much longer than broad .... *E. palustris*
               13. Tubercle depressed, not longer than broad .... *E. smallii*
            12. Spikelets with 1 basal empty scale, nearly encircling the base.
               14. Spikelet closely many-flowered, often with 40 or more opaque, fertile scales .... *E. calva*
               14. Spikelet open, few-flowered; fertile scales 5 to 30, lustrous ......
                  .... *E. uniglumis*

6. Style 3-cleft; achenes triangular.
    15. Plants annual, dwarf, more or less cespitose; stems capillary.
        16. Stems mostly erect; achenes 0.6 to 1 mm. long, light-gray to yellow .... *E. microcarpa*
        16. Stems weak, often reclining; achenes 1.5 mm. long, olive-brown .... *E. intermedia*
    15. Plants perennial, with rhizomes.
        17. Stems mostly about 1 dm. high, capillary or triangular, flaccid; style base jointed to the apex of the achene .... *E. acicularis*
        17. Stems 3 to 8 dm. long, compressed or angular, often rooting at the tip; style base confluent with the apex of the achene .... *E. rostellata*

Eleocharis acicularis (L.) R. and S. Fig. 66, A–C. Map 153.

Common in shallow water, partially dried ponds, muddy flats, and swales. Nearly throughout the United States. This is a variable species with several varieties.

Eleocharis calva Torr. Fig. 65, I–J. Map 154.

Abundant, usually forming dense mats along shores, in shallow water, and on sandy or marshy flats. Common in the northeastern states, widespread elsewhere.

Eleocharis caribaea (Rottb.) Blake. Fig. 64, H–I. Map 155.

In streams, ponds, ditches, and marshes. Mostly near the South Atlantic and Gulf Coasts, local in eastern Texas and California. Common in tropical America.

Eleocharis cellulosa Torr. Fig. 63, D–F. Map 156.

In shallow water and brackish marshes. Mostly along the Gulf Coast.

Eleocharis elongata Chapm. Fig. 63, G–H. Map 157.

Shallow lakes, ponds, and marshes. Florida to Texas.

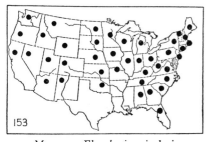

MAP 153. *Eleocharis acicularis.*

MAP 154. *Eleocharis calva.*

Map 155. *Eleocharis caribaea.*

Map 156. *Eleocharis cellulosa.*

Map 157. *Eleocharis elongata.*

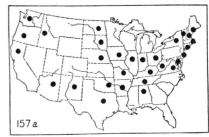

Map 157a. *Eleocharis engelmanni.*

Map 158. *Eleocharis equisetoides.*

Map 159. *Eleocharis intermedia.*

Eleocharis engelmanni Steud. Fig. 64, F–G. Map 157a.

Along wet shores and in shallow ponds and wet depressions. Widely distributed, chiefly in the eastern states.

Eleocharis equisetoides (Ell.) Torr. Fig. 63, A–C. Map 158.

In shallow water of ponds and slow streams; chiefly along the Coastal Plain. Massachusetts to Florida and Texas; local about the Great Lakes.

Eleocharis intermedia (Muhl.) Schult. (*E. reclinata* Kunth.). Fig. 66, G–H. Map 159.

In shallow ponds and wet depressions. New England to New Jersey, westward to Minnesota and Iowa.

Eleocharis microcarpa Torr. Fig. 66, I–J. Map 160.

In shallow water and on wet, muddy, or sandy shores. Chiefly along the Atlantic and Gulf Coasts.

Eleocharis obtusa (Willd.) Schult. Fig. 64, A–C. Map 161.

In muddy or sandy, partially desiccated ponds and swales. Widespread and common throughout the eastern United States and along the Pacific Coast.

Eleocharis olivacea Torr. Fig. 66, K–L. Map 160a.

On wet, sandy, or muddy shores, occasionally in shallow water. Maine to Florida and in the Great Lakes region.

Map 160. *Eleocharis microcarpa.*

Map 160a. *Eleocharis olivacea.*

Map 161. *Eleocharis obtusa.*

Map 161a. *Eleocharis ovata.*

Map 162. *Eleocharis palustris.*

Map 162a. *Eleocharis uniglumis.*

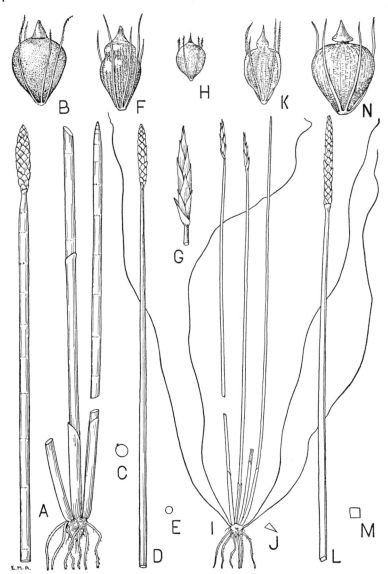

Fig. 63. *Eleocharis equisetoides* (A–C), *E. cellulosa* (D–F), *E. elongata* (G, H), *E. robbinsii* (I–K), and *E. quadrangulata* (L–N).

(A) Parts of plant showing fertile and sterile stems; x 1/2. (B, F, H, K, N) Achenes; x 8. (C, E, J, M) Diagrams of cross section of stem. (D) Fertile stem; x 1/2. (G) Spike; x 2. (I) Plant; x 1/2. (L) Stem terminated by spike; x 1/2.

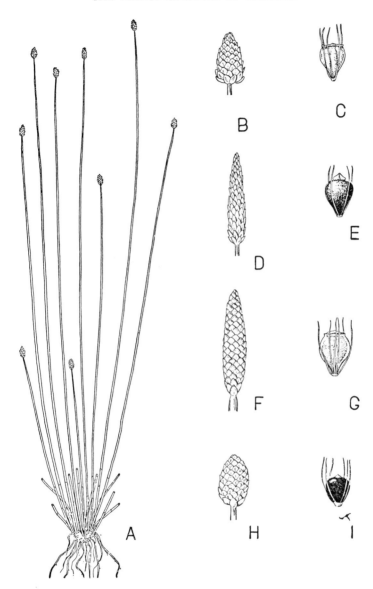

Fig. 64. *Eleocharis obtusa* (A–C), *E. ovata* (D, E), *E. engelmanni* (F, G), and *E. caribaea* (H, I).

(A) Plant; x 1/2. (B, D, F, H) Spikes; x 2. (C, E, G, I) Achenes; x 8.

Map 163. *Eleocharis quadrangulata*.   Map 164. *Eleocharis robbinsii*.

Eleocharis ovata R. and S. (*E. diandra* C. Wright). Fig. 64, D–E. Map 161a.

On sandy or muddy shores and on tidal flats of lakes, rivers, and estuaries. Maine to Pennsylvania; local about the Great Lakes and in Washington.

Eleocharis palustris (L.) R. and S. Fig. 65, A–D. Map 162.

Locally common in shallow water of lakes and streams and on adjacent marshes. Chiefly in the northern United States. Plants with stouter stems up to 5 mm. in diameter, var. *major* Sonder, occur in deep water.

Eleocharis quadrangulata (Michx.) R. and S. Fig. 63, L–N. Map 163.

In ponds and along sandy or peaty shores. Especially on the Coastal Plain and in the Mississippi Valley.

Eleocharis robbinsii Oakes. Fig. 63, I–K. Map 164.

In shallow water and on wet borders of ponds mostly in acid soils. Coastal Plain, from Maine to Florida, and about the Great Lakes.

Eleocharis rostellata Torr. Fig. 66, D–F. Map 165.

Salt marshes, ponds, and springs along the coast; marl bogs and springs inland; represented by a taller, more erect form, var. *occidentalis* Wats., in the western states.

Eleocharis smallii Britt. Fig. 65, E–F. Map 166.

In shallow water and along margins of ponds and streams; in sandy or peaty soils. Northeastern and north central states.

Map 165. *Eleocharis rostellata*.   Map 166. *Eleocharis smallii*.

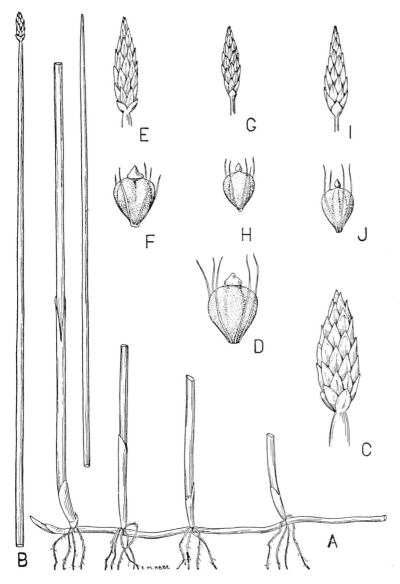

FIG. 65. *Eleocharis palustris* (A–D), *E. smallii* (E, F), *E. uniglumis* (G, H), and *E. calva* (I, J).

(A) Plant with rootstock; x 1/2. (B) Fertile stem; x 1/2. (C, E, G, I) Spikes; x 2. (D, F, H, J) Achenes; x 8.

158 AQUATIC PLANTS OF THE UNITED STATES

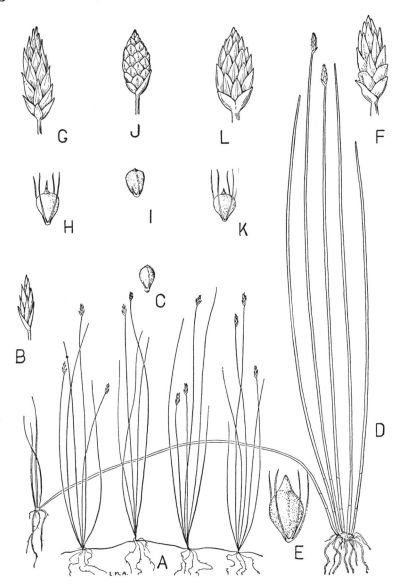

Fig. 66. *Eleocharis acicularis* (A–C), *E. rostellata* (D–F), *E. intermedia* (G, H), *E. microcarpa* (I, J), and *E. olivacea* (K, L).

(A) Plant showing creeping rootstock; x 1. (B) Spike; x 4. (C) Achene; x 8.
(D) Plant with runner; x 1/2. (F) Spike; x 2. (G, J, L) Spikes; x 4. (E, H, I, K) Achenes; x 8.

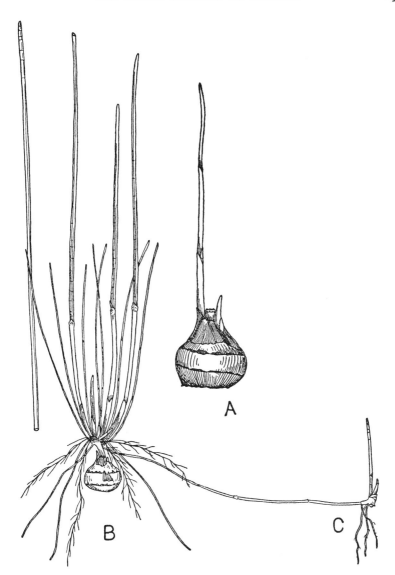

Fig. 67. *Eleocharis tuberosa.*
(A) Tuber; x 1/2. (B) New plant growing from tuber showing rhizome with new tuber beginning to form at C; x 1/4.

Eleocharis uniglumis (Link) Schult. Fig. 65, G–H. Map 162a.

In brackish, calcareous, or alkaline marshes, ponds, and adjacent shores. Mostly in New England and the Northwest.

Eleocharis tuberosa Schult. Water chestnut, Matai. Fig. 67.

NOTE. This is a species native to China where it is extensively grown under the name "matai" for its starchy, brown-coated tubers which are commonly used in Chinese cookery. The tubers are also used in the Chinese-American dish chop suey. In America it is often confused with the water nut, *Trapa natans,* the nuts of which are also called water chestnut (Fig. 123). *E. tuberosa* has recently been planted in the southern United States. It is yet too early to say whether the tubers can be produced in commercial quantities in this country. They are borne on the ends of slender rhizomes imbedded in the bottom mud or sand in shallow water or in areas subject to periodic inundation.

ERIOPHORUM: COTTON GRASS, WOOL GRASS

Perennials with flat, grasslike leaves, naked or leafy stems, and creeping rootstocks. Spikelets solitary or in naked or involucrate clusters. Scales of the spikelet 1- to 5-nerved; bristles numerous, simple, becoming long, exserted, and cottony in fruit; style slender, 3-cleft; achenes acutely triangular. — About 15 species, mostly of bogs in the Northern Hemisphere.

KEY TO SPECIES OF ERIOPHORUM

1. Cluster of spikelets like a head; scales striate; hairs tawny or white; achenes obovoid .... *E. virginicum*
1. Cluster of spikelets like an umbel; scales not striate; hairs white, rarely buff; achenes linear-oblong .... *E. viridi-carinatum*

MAP 167. *Eriophorum virginicum.*   MAP 168. *Eriophorum viridi-carinatum.*

Eriophorum virginicum L. Fig. 68, C–E. Map 167.

Common in acid bogs, floating moors, and shallow ponds. Eastern states and Great Lakes region.

Eriophorum viridi-carinatum (Engelm.) Fernald. Fig. 68, A–B. Map 168.

Northeastern states and Great Lakes region; rare in the northern Rocky Mountains.

RYNCHOSPORA: Beak Rush

Mostly perennials with slender, angular, leafy stems. Spikelets in terminal or axillary clusters, 1- or 2-flowered, ovate to spindle-shaped, terete or flattened, with several more or less open scales. The lower scales empty, the upper often with imperfect flowers. Stamens usually 3; pistil solitary, subtended by bristles. Achene lenticular, globular, or flat, surmounted by an elongated beak or tubercle. — About 200 species, mostly of warm regions; many of them occur in bogs and a few in shallow water.

KEY TO SPECIES OF RYNCHOSPORA

1. Style elongate; stigmas much shorter than the style; perianth bristles longer than the achene.
    2. Spikelets in 1 to 4 dense, globose heads; tubercle on achene short ...*R. traceyi*
    2. Spikelets in panicled clusters; tubercle about 3 times as long as the body of the achene.
        3. Plant stoloniferous; spikelets loosely spreading; bristles usually 4 *R. careyana*
        3. Plant not stoloniferous; spikelets fusiform, in fastigiate clusters; bristles usually 6 ....*R. macrostachya*
1. Style short; stigmas at least as long as the style.
    4. Bristles 9 to 20, much longer than the achene, retrorsely barbed ....*R. alba*
    4. Bristles 6, shorter than the achene, upwardly barbed ....*R. cymosa*

Rynchospora alba (L.) Vahl. Fig. 70, A–B. Map 171.

Common in wet bogs and shallow pools. Widespread in the eastern states and infrequent in the Pacific Northwest.

Rynchospora careyana Fernald. Fig. 70, E–F. Map 172.

In ponds, streams, and marshes in the sands of the Coastal Plain. Georgia and Florida.

Rynchospora cymosa Ell. Fig. 70, G–H. Map 173.

In bogs and swales. From Virginia to Texas.

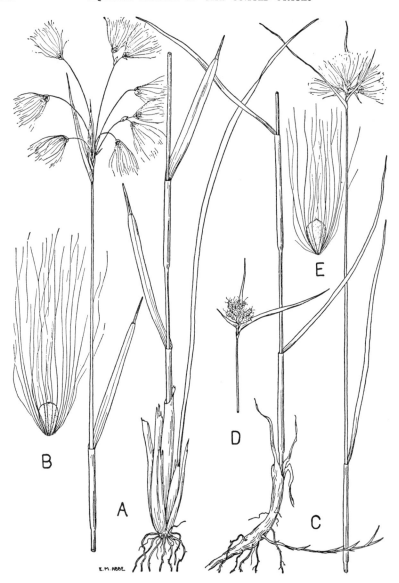

Fig. 68. *Eriophorum viridi-carinatum* (A, B) and *E. virginicum* (C-E).

(A) Plant in fruiting state; x 1/2. (B) Achene with bristles; x 3.
(C) Plant in fruiting state; x 1/2. (D) Cluster of spikelets in flower; x 1/2. (E) Achene with bristles; x 3.

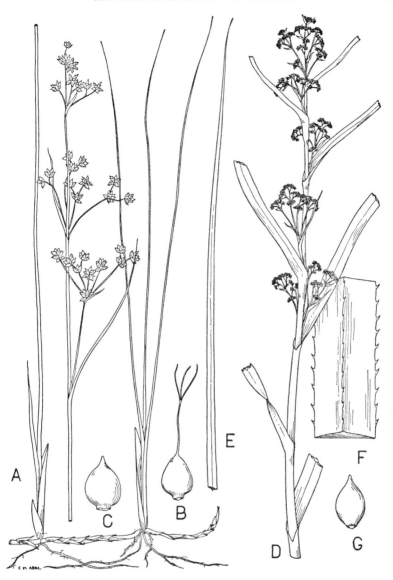

Fig. 69. *Cladium mariscoides* (A–C) and *C. jamaicensis* (D–G).

(A) Plant; x 1/2. (B) Pistil; x 5. (C) Achene; x 5.
(D) Upper part of flowering plant with leaves cut off; x 1/2. (E) Upper part of a leaf; x 1/2. (F) Small section of a leaf, lower surface turned up to show sharp spines; x 5. (G) Achene; x 5.

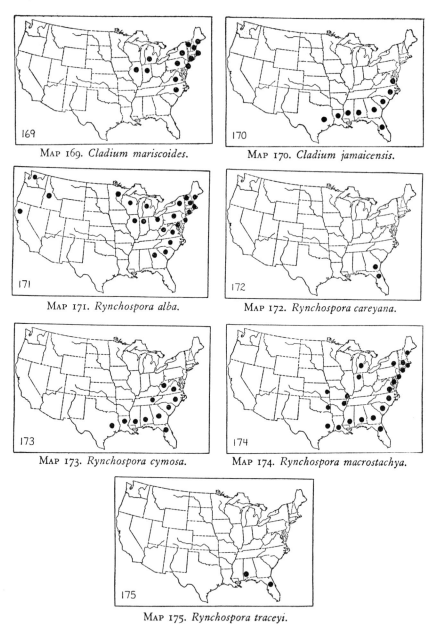

Map 169. *Cladium mariscoides.*
Map 170. *Cladium jamaicensis.*
Map 171. *Rynchospora alba.*
Map 172. *Rynchospora careyana.*
Map 173. *Rynchospora cymosa.*
Map 174. *Rynchospora macrostachya.*
Map 175. *Rynchospora traceyi.*

Rynchospora macrostachya Torr. Fig. 70, I–J. Map 174.

In marshes, shallow ponds, and slow streams. Common along the Atlantic and Gulf Coasts; local in the Mississippi Valley and Great Lakes region.

Rynchospora traceyi Britt. Fig. 70, C–D. Map 175.
Local in sandy ponds and wet depressions on the Coastal Plain. Florida and Alabama.

Fig. 70. *Rynchospora alba* (A, B), *R. traceyi* (C, D), *R. careyana* (E, F), *R. cymosa* (G, H), and *R. macrostachya* (I, J).
(A, C, E, G, I) Habit showing fruiting stage; x 1/2. (B, D, F, H, J) Achenes; x 5.

SCIRPUS: BULRUSH, CLUB RUSH

Stems triangular or terete, leafy or naked, except for the basal sheaths. Spikelets rarely solitary, mostly in terminal clusters subtended by an involucre of 1 or several leafy bracts, all but the lowest 1 or 2 bearing flowers. Flowers perfect, with 2 or 3 stamens and an ovary with 2- or 3-cleft style; bristles 1 to 8 or wanting. Achenes lenticular, plano-convex, or triangular. — About 150 species; widely distributed. Many of them grow in ponds and lakes, others in wet swales and marshes.

KEY TO SPECIES OF SCIRPUS

1. Stems triangular, from creeping rootstocks.
    2. Involucral bracts several; spikelets often on long rays; rootstocks with series of tuberous thickenings.
        3. Achenes sharply triangular .... S. fluviatilis
        3. Achenes plano-convex or obscurely triangular.
            4. Scales of spikelets with elongate, reddish markings .... S. robustus
            4. Scales of spikelets without reddish markings .... S. paludosus
    2. Involucral bract solitary; rootstocks without tubers.
        5. Spikelets in open, umbel-like corymb.
            6. Bract terete, short, like an extension of the stem .... S. californicus
            6. Bract flat, long, leaflike .... S. etuberculatus
        5. Spikelets in a dense cluster, rarely solitary.
            7. Bract 1 to 3 cm. long .... S. olneyi
            7. Bract 4 to 16 cm. long.
                8. Achene plano-convex; scales reddish-brown; bristles shorter than the achene; leaves mostly not half as long as the stem .... S. americanus
                8. Achene triangular; scales yellow-brown; bristles longer than the achene; leaves more than half as long as the stem .... S. torreyi
1. Stems terete or nearly so; involucral bract appearing as an extension of the stem.
    9. Spikelets solitary; stems slender, not rigid; leaves basal, linear, flaccid; rhizomes slender, tuber-bearing .... S. subterminalis
    9. Spikelets in clusters; stems firm.
        10. Plants annual, small, and slender; spikelets sessile or in dense clusters .. .. S. smithii
        10. Plants perennial, with coarse stems 1 to 4 meters high, from creeping rootstocks; spikelets in open umbel-like or paniculate clusters.
            11. Stems triangular toward apex .... S. californicus
            11. Stems terete all along.
                12. Achenes triangular .... S. heterochaetus
                12. Achenes not triangular.
                    13. Stems light green, soft; achene 2 mm. long; spikelets ovate, spreading .... S. validus
                    13. Stems dark green, firm; achene 2.5 to 3 mm. long; spikelets lanceolate, close .... S. acutus

Scirpus acutus Muhl. Tule, Hard-stem bulrush. Fig. 74, E–G. Map 176.

In shallow ponds and lakes, also in sloughs and marshes. In calcareous regions this species often covers extensive areas, the so-called "tule marshes." Widespread throughout the United States except in the Southeast.

Scirpus americanus Pers. Three-square, Shore rush. Fig. 71, D–E. Map 177.

In shallow water or on sandy shores of lakes, ponds, and streams and along the seashore; most abundant in saline, brackish, or alkaline waters. Widespread.

Scirpus californicus (Meyer) Britt. Map 178.

In ponds, marshes, and springy places. Local in the southern states and California.

Scirpus etuberculatus (Steud.) Ktze. Map 179.

In ponds, marshes, and swamps. From Maryland southward, mostly near the coast.

Scirpus fluviatilis (Torr.) Gray. River bulrush. Fig. 72. Map 179.

In sloughs and borders of ponds and bays of larger lakes. Widespread, chiefly in the northern states.

Map 176. *Scirpus acutus.*

Map 177. *Scirpus americanus.*

Map 178. *Scirpus californicus.*

Map 179. *Scirpus etuberculatus,* +; *S. fluviatilis,* ●.

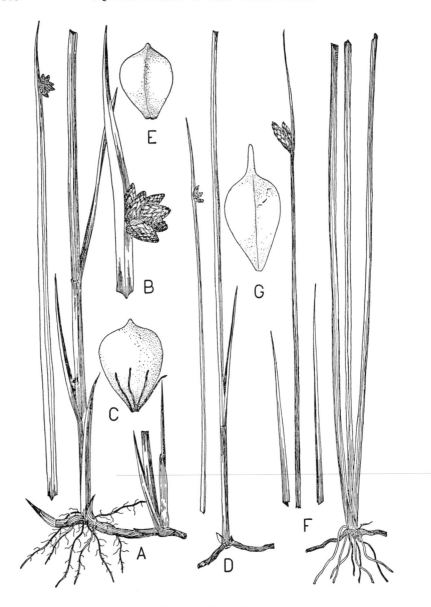

Fig. 71. *Scirpus olneyi* (A–C), *S. americanus* (D, E), and *S. torreyi* (F, G).
(A) Plant; x 1/2. (B) Cluster of spikelets; x 2. (C) Achene; x 10.
(D) Plant; x 1/2. (E) Achene; x 10.
(F) Plant; x 1/2. (G) Achene; x 10.

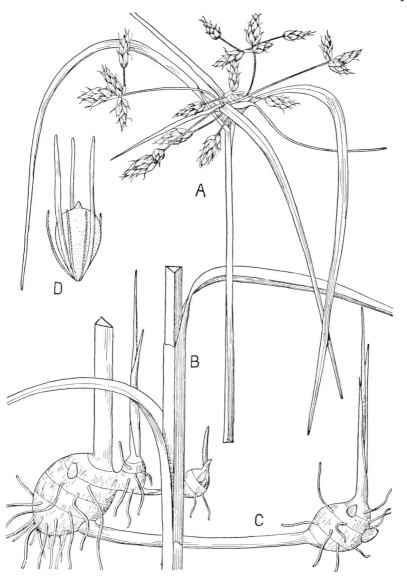

FIG. 72. *Scirpus fluviatilis*.

(A) Cluster of fruiting spikelets; x 1. (B) Triangular stem with closed leaf sheath; x 1. (C) Base of plant showing rhizome and tubers; x 1. (D) Achene; x 5.

Map 180. *Scirpus heterochaetus.*

Map 181. *Scirpus olneyi.*

Map 182. *Scirpus paludosus.*

Map 183. *Scirpus robustus.*

Scirpus heterochaetus Chase. Fig. 74, H–I. Map 180.

In marshes and shallow water of lakes and ponds. Common in the north central states; infrequent elsewhere.

Scirpus olneyi Gray. Fig. 71, A–C. Map 181.

In shallow water of salt or brackish pools and in marshes and beaches along the seashore. Mostly near the Atlantic and Gulf Coasts and inland in the western states.

Scirpus paludosus A. Nels. Fig. 73, A–C. Map 182.

In brackish or alkaline ponds and marshes. Common west of the Mississippi River, local in the northeastern states.

Scirpus robustus Pursh. Fig. 73, D–E. Map 183.

In brackish or saline ponds and marshes. Chiefly along the Atlantic and Gulf Coasts; rare inland.

Scirpus smithii Gray. Fig. 75, E–H. Map 184.

In shallow water and on sandy or muddy flats along streams and lakes. Most common in the eastern states.

Scirpus subterminalis Torr. Water club rush. Fig. 75, A–D. Map 185.

In sluggish streams and shallow ponds and lakes. Common in New England and northern New York, local about the Great Lakes and Pacific Northwest.

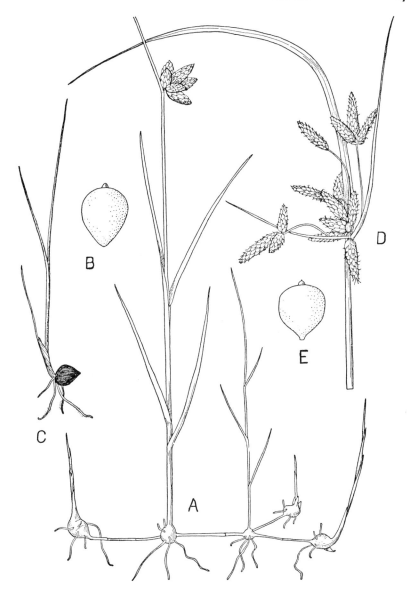

Fig. 73. *Scirpus paludosus* (A–C) and *S. robustus* (D, E).

(A) Habit of plant showing rhizomes and tubers; x 1/2. (B) Achene; x 5. (C) Seedling; x 2.
(D) Cluster of spikelets; x 1/2. (E) Achene; x 5.

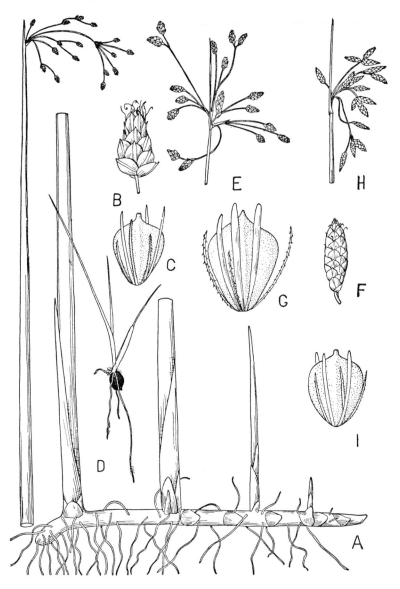

Fig. 74. *Scirpus validus* (A–D), *S. acutus* (E–G), and *S. heterochaetus* (H, I).

(A) Habit of plant with rhizome; x 1/2. (B) Spikelet; x 3. (C) Achene; x 10. (D) Seedling; x 2.

(E) Cluster of spikelets; x 1/2. (F) Spikelet; x 2. (G) Achene; x 10.

(H) Cluster of spikelets; x 1/2. (I) Achene; x 10.

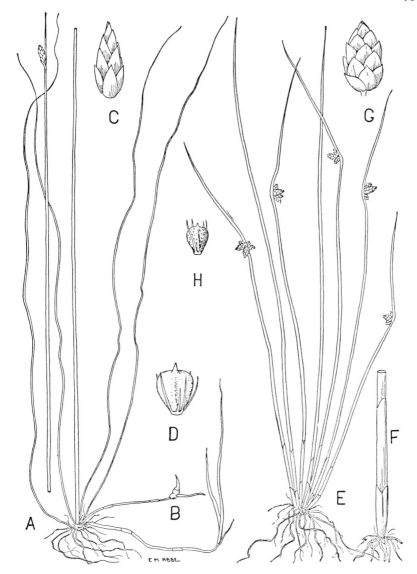

FIG. 75. *Scirpus subterminalis* (A–D) and *S. smithii* (E–H).

(A) Plant showing habit; x 1/2. (B) Tuber; x 1/2. (C) Spikelet; x 3. (D) Achene; x 5.
(E) Plant showing habit; x 1/2. (F) Base of stem showing sheaths; x 2. (G) Spikelet; x 3. (H) Achene; x 5.

Scirpus torreyi Olney. Fig. 71, F–G. Map 186.

In shallow ponds and marshes, usually in colder water. Mostly from New England to Minnesota.

Scirpus validus Vahl. Great bulrush, Soft-stem bulrush. Fig. 74, A–D. Map 187.

In shallow ponds, especially about spring holes, bog holes, and stream banks. Widespread throughout the United States.

Map 184. *Scirpus smithii.*

Map 185. *Scirpus subterminalis.*

Map 186. *Scirpus torreyi.*

Map 187. *Scirpus validus.*

REFERENCES

Svenson, H. K. Monographic studies in the genus *Eleocharis*. Rhodora 31:121–135, 152–163, 167–191, 199–219, 224–242. 1929. 34:193–203, 215–227. 1932. 36:377–389. 1934. 39:210–231, 236–273. 1937. 41:1–19, 43–77, 90–110. 1939.

Fernald, M. L., and A. E. Brackett. The representatives of *Eleocharis palustris* in North America. Rhodora 31:57–77. 1929.

Fernald, M. L. Critical notes on *Carex*. Rhodora 44:281–331. 1942.

Beetle, A. A. Studies in the genus *Scirpus* L. Amer. Jour. Bot. 27:63–64. 1940. 28:469–476, 691–708. 1941. 29:82–88, 653–656. 1942. 30:395–401. 1943.

———. A key to the North American species of the genus *Scirpus* based on achene characters. Amer. Midl. Nat. 29:533–538. 1943.

Mackenzie, K. K. North American Cariceae. 2 vols. 1940.

———. *Carex*. In, North Amer. Flora. 18:1–478. 1931–1935.

ARACEAE: Arum Family

Stout plants with alternate, simple or compound, fleshy leaves. Flowers reduced, often without perianth, imperfect or perfect, crowded on a spadix which in most genera is surrounded by a large bract, the spathe. Fruit a berry or utricle. — This is a large family composed chiefly of terrestrial forms reaching their greatest development in tropical regions. Several genera contain a few aquatic or semiaquatic species.

KEY TO GENERA

1. Plant floating; leaves prominently ridged or plaited, in dense rosettes; spathe minute, white .... *Pistia*
1. Plant anchored by roots (rarely floating in *Calla*).
   2. Leaves alternate; rootstock stout, creeping.
      3. Leaves cordate, petioled; spadix terminal, nearly globular; spathe white .... *Calla*
      3. Leaves swordlike, sheathing; spadix cylindric, lateral on a leaflike scape; spathe wanting .... *Acorus*
   2. Leaves in a basal cluster; rootstock short, erect.
      4. Leaves lanceolate, with parallel venation; spadix cylindrical, yellow; spathe wanting .... *Orontium*
      4. Leaves sagittate to cordate, with pinnate venation; spadix ovoid, green; spathe fleshy, green .... *Peltandra*

ACORUS: Sweet Flag

Perennials with stout, creeping, aromatic rootstocks and long, swordlike, 2-ranked leaves. Spadix cylindrical, yellow-green, appearing lateral on a leaflike tapering scape. Flowers with 6 perianth parts and 6 stamens; ovary 2- to 3-celled, with several ovules; fruits with 1 to 4 seeds, cuneate, dry but gelatinous inside.

Acorus calamus L. Sweet flag. Fig. 76, C–D. Map 188.

Widespread; in margins of shallow ponds, streams, and sloughs and in spring-fed marshes.

Note. *A. calamus* may persist and even spread over considerable areas in partially drained marshy fields which become relatively dry in summer. In such a habitat it may not flower or fruit for years, undoubtedly accounting for the belief prevailing in some localities that it does not fruit in America. Plants growing in shallow water or in spring-fed marshes usually fruit and produce viable seeds in abundance (Fig. 77, A–B).

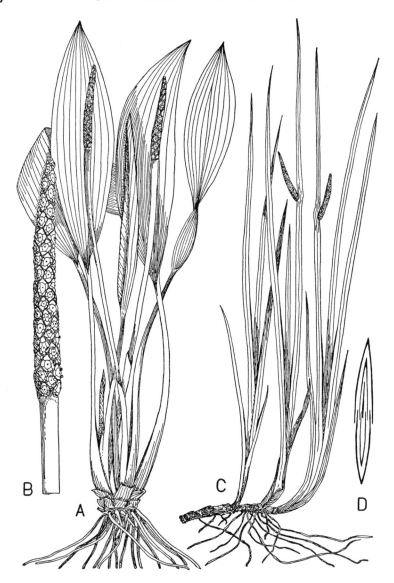

FIG. 76. *Orontium aquaticum* (A, B) and *Acorus calamus* (C, D).

(A) Habit sketch of plant with inflorescences; x 1/4. (B) Naked spadix; x 1.
(C) Habit sketch of plant in flower; x 1/6. (D) Cross section diagram of leaf arrangement.

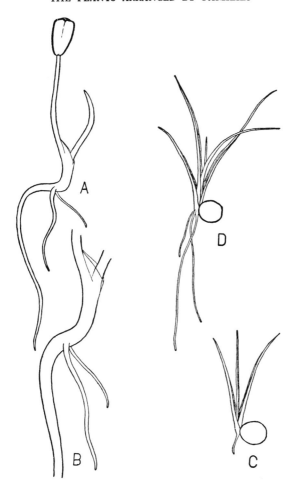

Fig. 77. *Acorus calamus* (A, B) and *Orontium aquaticum* (C, D).

(A) Seedling about 2 weeks after seed germination; x 4. (B) Base of same showing first adventitious roots; x 6.

(C, D) Seedlings 1 and 2 weeks after seed germination; x 1.

CALLA: WATER ARUM

Perennials with stout, creeping rhizomes and alternate, petioled, cordate leaves. Spadix short-cylindric to ovate, covered all over with naked flowers, the lower perfect and the upper often staminate; spathe ovate, about 2 to 5 cm. long, white on the upper surface; berries red, with a few brown, cylindrical seeds.

Map 188. *Acorus calamus.*   Map 189. *Calla palustris.*

Calla palustris L. Wild calla. Fig. 78. Map 189.

Along margins of cold, spring-fed ponds and brooks and in bogs, rooted or at times at least partly floating on the surface. Northeastern states and Great Lakes region.

ORONTIUM: Golden Club

Perennials with short, erect, deep-seated rootstocks and numerous fibrous roots. Leaves in a cluster, petioled, lanceolate, entire, the blade 1 to 3 dm. long, frequently floating. Spadix about 1 to 2 dm. long, the lower flowers with 6 perianth parts and 6 stamens, the upper with 4. Pistil solitary; ovary with 1 ovule; fruit a green utricle with a large seed about 6 to 10 mm. in diameter.

Orontium aquaticum L. Golden club. Figs. 76, A–B; 77, C–D. Map 190.

In shallow ponds, swamps, and tidal flats of rivers. Eastern and southern states.

PELTANDRA: Arrow Arum

Stout perennials with short, erect rootstocks and basal, fleshy, sagittate to cordate, rarely entire leaves. Spathe fleshy, green, convolute below and often spreading above, with wavy margin. Spadix covered by imperfect, naked flowers or sterile above. Staminate flowers above, of 6 to 10 stamens. Pistillate flowers below, of 1-celled carpels surrounded by 4 or 5 staminodia. Fruits fleshy, greenish or purplish, 1 to 2 cm. in diameter, with 1 to 3 seeds imbedded in a gelatinous mass. Seeds without endosperm.

KEY TO SPECIES OF PELTANDRA

1. Spathes narrow, green; fruits green ....*P. virginica*
1. Spathes spreading above, whitish; fruits reddish ....*P. sagittaefolia*

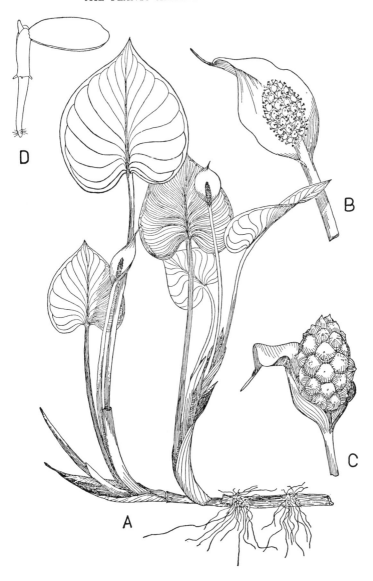

Fig. 78. *Calla palustris*.

(A) Plant; x 1/4. (B) Spadix with flowers; x 1. (C) Spadix in fruit; x 1. (D) Young seedling; x 5.

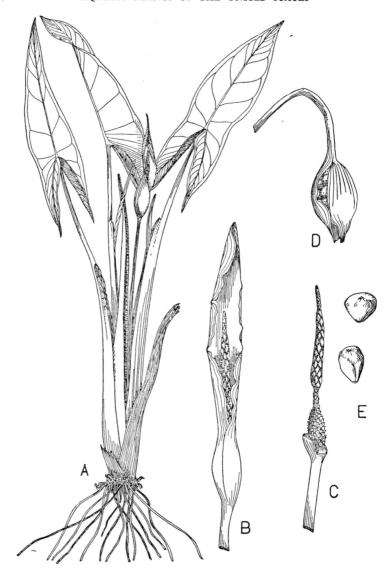

Fig. 79. *Peltandra virginica*.

(A) Plant; x 1/4. (B) Spadix in spathe; x 1/2. (C) Spadix with spathe removed to show staminate flowers above and pistillate flowers below; x 1/2. (D) Mature fruits in spathe; x 1/2. (E) Fleshy, 1-seeded fruits; x 1.

Peltandra virginica (L.) Kunth. Arrow arum. Figs. 79; 90, B. Map 191.

In shallow water and sloughs and along banks of streams. Mostly in the eastern states.

Peltandra sagittaefolia (Michx.) Morong.

In bogs and springy places. In the southeastern states.

Map 190. *Orontium aquaticum*.   Map 191. *Peltandra virginica*.

PISTIA: Water Lettuce

Plants mostly floating or becoming stranded; leaves sessile, plaited, fleshy, in rosettes on short stems bearing numerous adventitious branching roots. Vegetative propagation by buds. Flowers imperfect, naked; on a small spadix which is fused to the median line of the spathe. Spathe whitish, about 1 cm. long, slightly constricted near the middle forming 2 cavities, the lower containing a pistil and the upper a whorl of 3 to 8 stamens whose filaments are fused into a column for the entire length making the anthers appear to terminate the spadix. Pistil 1-celled, with a few ovules of which usually only one produces a seed.

Pistia stratiotes L. Water lettuce. Fig. 80. Map 192.

Floating on the surface of sluggish ponds, streams, and drainage ditches. Chiefly in Florida and along the Gulf Coast. Common in the tropics where sometimes it completely covers the water surface. It is not hardy northward but frequently it is used in conservatory and garden pools.

REFERENCE

Buell, M. F. *Acorus calamus* in America. Rhodora 37:367-369. 1935.

LEMNACEAE: Duckweed Family

Plants minute, free-floating or submersed, with undifferentiated, flattened or globular plant body (frond) without definite leaf or stem, single or aggregated in small colonies. Fronds with or without roots,

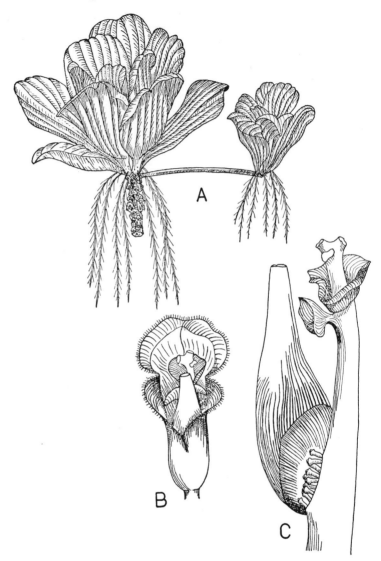

FIG. 80. *Pistia stratiotes*.

(A) Habit sketch of plant showing vegetative propagation; x 1/4. (B) Front view of inflorescence; x 2. (C) Side view of inflorescence; x 4.

Map 192. *Pistia stratiotes*.

propagating readily by buds bursting through a flap of the vegetative pouches. Flowers borne in a reproductive pouch on the margin or upper surface of the frond, imperfect, much reduced, naked, usually appearing only after a period of hot weather; fruit a utricle. — The family contains 4 genera of temperate and tropical regions.

KEY TO GENERA

1. Fronds without roots; propagative pouches solitary.
   2. The frond thin, sickle-shaped or much elongated .... *Wolffiella*
   2. The frond thick, globular or ellipsoidal .... *Wolffia*
1. Fronds with roots from the lower surface; propagative pouches 2.
   3. Roots solitary on each frond .... *Lemna*
   3. Roots 2 or more on each frond .... *Spirodela*

LEMNA: Duckweed

Fronds flattened, 1- to 5-nerved, proliferating from 2 vegetative pouches on the margin near the base, usually bearing a single root. Flowers 3, borne in the flowering pouch, 2 staminate, each composed of 1 stamen, and the other pistillate, composed of a single naked pistil. Ovary 1-celled, with 1 to several ovules. Fruit usually with 1 seed.

KEY TO SPECIES OF LEMNA

1. Fronds oblong with tapering base, stalked, 6 to 12 mm. long, often forming extensive submersed colonies .... *L. trisulca*
1. Fronds oblong-ovate to elliptical, not stalked, 2 to 5 mm. long, floating.
   2. Outline of frond symmetrical or nearly so.
      3. Frond oblong-elliptical .... *L. minima*
      3. Frond oblong-obovate.
         4. Upper surface uniformly green .... *L. minor*
         4. Upper surface mottled or irregularly brown-streaked .... *L. gibba*
   2. Outline of frond oblique.
      5. Frond narrow, obscurely 1-nerved .... *L. ~~cyclostasa~~* valdiviana
      5. Frond obliquely obovate, 3- to 5-nerved.
         6. Lower surface of frond convex .... *L. gibba*
         6. Lower surface of frond flat .... *L. perpusilla*

Lemna cyclostasa (Ell.) Chev. (*L. valdiviana* Phill.). Fig. 81, H. Map 193.

In cold, spring-fed ponds and sluggish streams. Chiefly in the eastern states.

Lemna gibba L. Fig. 81, G.

In stagnant ponds and pools. Widespread but local; Nebraska to Texas and California.

Lemna minima Phill. Map 194.

In pools and ponds. Widespread but local.

Lemna minor L. Fig. 81, A–F. Map 195.

In stagnant ponds and sluggish streams. A common species throughout the United States.

Lemna perpusilla Torr. Map 196.

Ponds and slow streams. Chiefly in the central states.

Lemna trisulca L. Fig. 82, L. Map 197.

In cold, spring-fed ponds, lakes, and brooks. Widely distributed except in the southern states.

Map 193. *Lemna cyclostasa.*

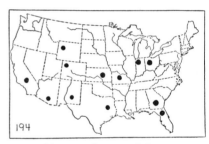

Map 194. *Lemna minima.*

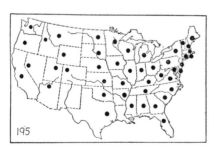

Map 195. *Lemna minor.*

Map 196. *Lemna perpusilla.*

Map 197. *Lemna trisulca.*

Map 198. *Spirodela polyrhiza,* ●; *S. oligorhiza,* ▲.

Map 199. *Wolffia columbiana.*

Map 200. *Wolffia punctata.*

SPIRODELA: Great Duckweed

Fronds flattened, 5- to 15-nerved, bearing several roots, similar to those of *Lemna,* but usually larger and often purple underneath; vegetative pouches 2. Flowers as in *Lemna,* 2 or 3 staminate and 1 pistillate in a pouch.

KEY TO SPECIES OF SPIRODELA

1. Fronds with 5 to 11 lateral veins .... *S. polyrhiza*
1. Fronds with 2 to 4 lateral veins .... *S. oligorhiza*

Spirodela oligorhiza Kurz. Fig. 82, B. Map 198.
  In ponds. Reported from Missouri.

Spirodela polyrhiza (L.) Schleid. Large duckweed. Fig. 82, A. Map 198.
  Common in stagnant ponds and sluggish streams. Throughout the United States.

WOLFFIA

Fronds thick, globular, ellipsoid, rootless, proliferating from a single, funnel-shaped pouch but soon separating, about 1 to 2 mm. long.

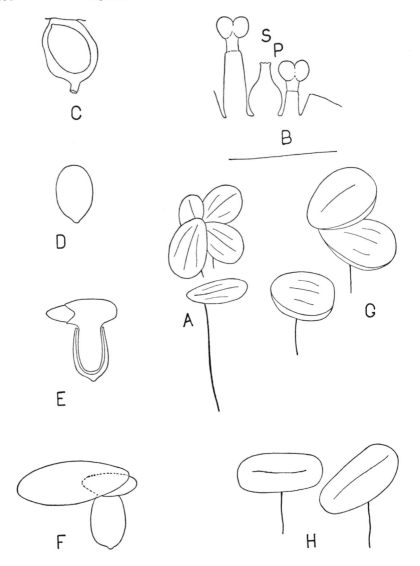

FIG. 81. *Lemna minor* (A-F), *L. gibba* (G), and *L. cyclostasa* (H).

(A) Plants showing habit; x 5. (B) Flowers (S, staminate; P, pistillate); x 15. (C) Fruit; x 15. (D) Seed; x 15. (E) Germinating seed; x 15. (F) Seedling; x 15.
(G) Plants; x 8.
(H) Plants; x 8.

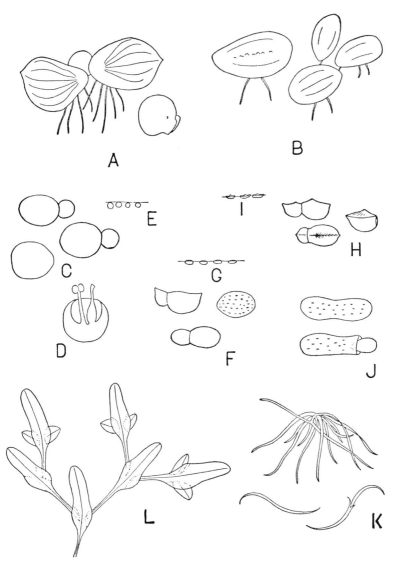

FIG. 82. Lemnaceae. (A) *Spirodela polyrhiza*, vegetative plants and "winter bud"; x 4. (B) *S. oligorhiza;* x 4. (C) *Wolffia columbiana*, vegetative plants; x 12. (D) Flowering plant with stamen and pistil; x 12. (E) Diagram showing position of plants just below surface of water. (F) *W. punctata;* x 12. (G) Diagram showing plants projecting above surface of water. (H) *W. papulifera;* x 12. (I) Diagram showing plants projecting above surface of water. (J) *Wolffiella lingulata*, vegetative and budding plants; x 3. (K) *W. floridana*, cluster of plants and single plant showing budding; x 3. (L) *Lemna trisulca;* x 3.

Flowers 2 in a pouch from the upper surface of the frond, the staminate a single stamen and the pistillate a single pistil.

NOTE. The entire plant has the appearance of a tiny green pea about the size of a pinhead and is the smallest seed plant known. While these plants normally grow in water, they frequently thrive among the damp vegetable matter on the bottom of a dried-up pond or slough. They have been found in an active vegetative condition by removing a layer of broken leaves and stems of sedges a decimeter deep in a dried-up pond.

KEY TO SPECIES OF WOLFFIA

1. Frond globose, light green all over, upper surface strongly convex, not dotted, floating under water; stomata 1 to 6 .... *W. columbiana*
1. Frond elongate, upper surface brown-dotted, floating on the surface of the water; stomata numerous.
    2. Upper surface rounded, with a prominent papilla in the center .. *W. papulifera*
    2. Upper surface flattened, without papilla .... *W. punctata*

Wolffia columbiana Karst. Fig. 82, C–E. Map 199.

Common in ponds, sloughs, and sluggish streams. Throughout the eastern and central United States.

Wolffia papulifera Thompson. Fig. 82, H–I.

In ponds and pools. Infrequent in the central states.

Wolffia punctata Griseb. Fig. 82, F–G. Map 200.

Locally common in ponds and sloughs; frequently associated with *W. columbiana* and *Lemna minor*. Widespread, chiefly in the eastern United States.

WOLFFIELLA

Fronds rootless, sickle-shaped, or strap-shaped, flattened and tapering on the ends, often several clinging in star-shaped colonies; proliferation by a bud from a single basal pouch. Flowers 2 in a pouch, the staminate a single stamen and the pistillate a single pistil.

KEY TO SPECIES OF WOLFFIELLA

1. Frond sickle-shaped.
    2. Length 10 to 20 times the width .... *W. floridana*
    2. Length 3 to 6 times the width .... *W. oblonga*
1. Frond straight, 1 to 4 times as long as wide .... *W. lingulata*

Wolffiella floridana (Smith) Thompson. Fig. 82, K. Map 201.
In ponds, ditches, and sluggish streams, usually intermixed with *Lemna*. Chiefly in the southeastern states.

Wolffiella lingulata Hegelm. Fig. 82, J. Map 201a.
In stagnant pools and ditches. Southern California to Mexico.

Wolffiella oblonga (Phill.) Hegelm. Map 201a.
In ponds and sluggish streams. Southern California and Mexico.

Map 201. *Wolffiella floridana*.

Map 201a. *Wolffiella oblonga*, ●; *W. lingulata*, ▲.

REFERENCES

Hegelmaier, F. Systematische Uebersicht der Lemnaceen. Engler, Bot. Jahrb. 21:3, 268–305. 1895.
Thompson, C. H. A revision of the Lemnaceae occurring north of Mexico. Rept. Mo. Bot. Gard. 9:21–42. 1898.
Goebel, K. Zur Organographie der Lemnaceen. Flora 114:278–305. 1921.
Saeger, Albert. The flowering of Lemnaceae. Torrey Bot. Club Bull. 56:351–358. 1929.
Hicks, L. E. Flower production in the Lemnaceae. Ohio Jour. Sci. 32:115–131. 1932.
Gilbert, H. C. Lemnaceae in flower. Sci. 86:308. 1937.
Mason, H. L. The flowering of *Wolffiella lingulata* (Hegelm.) Hegelm. Madroño 4:241–251. 1938.
Blake, C. H. *Wolffiella floridana* in Massachusetts. Rhodora 40: 76. 1938.
Brooks, J. S. The cytology and morphology of the Lemnaceae. Cornell Univ. Thesis. 1940.

MAYACACEAE: Bog-Moss Family

Small, mosslike, submersed herbs with branched, creeping or floating stems. Leaves crowded, linear, sessile, pellucid, entire, and notched at apex. Flowers solitary on axillary peduncles, perfect, and regular. Fruit a capsule; seeds several, with endosperm. — A single genus, *Mayaca*.

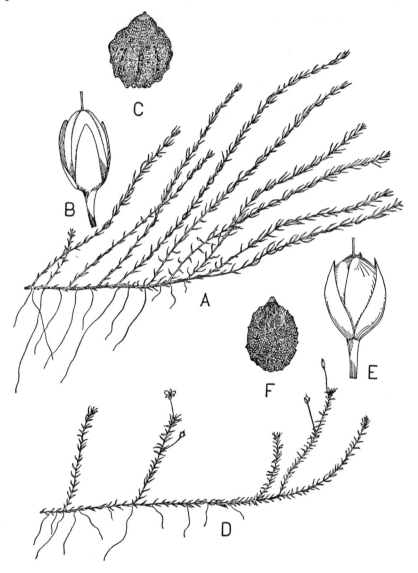

Fig. 83. *Mayaca fluviatilis* (A–C) and *M. aubleti* (D–F).
(A) Plant; x 1/2. (B) Capsule; x 5. (C) Seed; x 20.
(D) Plant; x 1/2. (E) Capsule; x 5. (F) Seed; x 20.

MAYACA: Pool Mosses, Bog Mosses

Flowers small, with persistent calyx of 3 herbaceous, lanceolate sepals and corolla of 3 obovate, white, pink, or violet petals. Stamens 3, alternate with the petals; pistil with 1-celled, superior ovary and filiform style; capsule 3-sided, with 3 parietal placentae and 6 to 12 ovoid or globose seeds. — About 6 species, mostly of warmer regions.

KEY TO SPECIES OF MAYACA

1. Stems elongate, trailing, up to 40 cm. long; pedicels shorter than the leaves .. ..M. fluviatilis
1. Stems tufted or matted, 2 to 20 cm. long; pedicels mostly exceeding the leaves ....M. aubleti

Mayaca aubleti Michx. Fig. 83, D–F. Map 202.

Local in shallow water and on margins of ponds, bogs, and spring-fed streams. Chiefly in the southeastern states.

Mayaca fluviatilis Aubl. Fig. 83, A–C. Map 203.

In shallow water of ponds and sluggish streams. In the southeastern states.

Map 202. *Mayaca aubleti.*   Map 203. *Mayaca fluviatilis.*

ERIOCAULACEAE: Pipewort Family

Monoecious or, rarely, dioecious, acaulescent plants with a rosette of awl-shaped leaves and a tuft of fibrous roots from a very short stem or axis. Scapes usually several, erect, slender, naked, with basal sheath, and terminating in a dense involucrate head of small flowers. The flowers axillary in scarious bracts, regular, imperfect. Stamens 2 to 6; ovary superior, 2- or 3-celled, with terminal style and 2 or 3 stigmas. Fruit a small capsule with 2 or 3 seeds with endosperm. — Several

ERIOCAULON: Pipewort, Button Rods

Perennials with very short stem covered with a rosette of flat or concave, glabrous, loosely cellular, often translucent leaves from 2 to 10 cm. long. Roots white, septate, often translucent. Heads depressed or convex, with staminate or pistillate flowers only, with both kinds intermixed, or with the staminate in the center and the pistillate on the outside. The flowers and the tips of the bracts often white-woolly or bearded. Staminate flowers with calyx of 2- or 3-keeled sepals and tubular corolla with 2 or 3 lobes; stamens 4 to 6 (rarely 3). Pistillate flowers with calyx of 2 or 3 sepals often attached some distance below the corolla; corolla of 2 or 3 separate narrow petals, or rarely wanting; stamens none; pistil with stalked or sessile ovary and 2 or 3 slender stigmas. — This genus contains many species of dry or wet sandy places and a few which are true aquatics or semiaquatics.

KEY TO SPECIES OF ERIOCAULON

1. Mature heads white-villous on top.
    2. Bractlets of receptacle longer than the flower .... *E. decangulare*
    2. Bractlets of receptacle not exceeding the flower.
        3. Petals of staminate flowers very unequal in size; plants mostly dioecious .... *E. compressum*
        3. Petals of staminate flowers equal or nearly so; plants monoecious .... *E. septangulare*
1. Mature heads glabrous or nearly so, mostly gray, black, or straw-colored.
    4. Bractlets of involucre chaffy, lustrous, spreading or reflexed ...... *E. ravenelii*
    4. Bractlets of involucre not chaffy or lustrous, appressed .... *E. parkeri*

Eriocaulon compressum Lam. Fig. 84, K–L. Map 205.

In shallow ponds, streams, and wet depressions in the pine barrens. Along the coast from New Jersey to Texas.

Eriocaulon decangulare L. Fig. 84, A–D. Map 204.

In shallow ponds and swamps. Chiefly in sandy soil of the pine barrens along the coast from New Jersey to Texas.

Eriocaulon parkeri Robinson. Fig. 84, E–G. Map 206.

Local on tidal mud flats of rivers and estuaries. New England to Virginia.

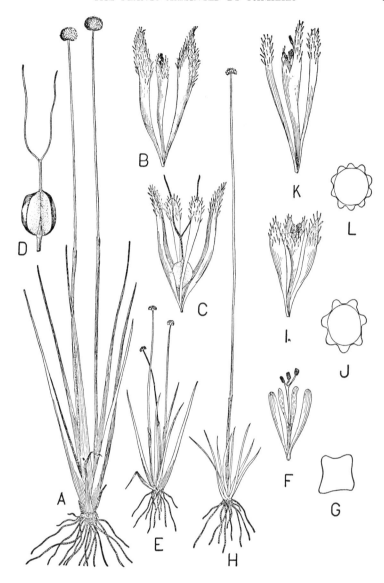

Fig. 84. *Eriocaulon decangulare* (A–D), *E. parkeri* (E–G), *E. septangulare* (H–J), and *E. compressum* (K, L).

(A) Plant; x 1/2. (B) Staminate flower; x 15. (C) Pistillate flower; x 15. (D) Fruit; x 20.
(E) Plant; x 1/2. (F) Staminate flower; x 15. (G) Cross section of peduncle; x 15.
(H) Plant; x 1/2. (I) Staminate flower; x 15. (J) Cross section of peduncle; x 15.
(K) Staminate flower; x 15. (L) Cross section of peduncle; x 15.

Map 204. *Eriocaulon decangulare.*

Map 205. *Eriocaulon compressum.*

Map 206. *Eriocaulon parkeri.*

Map 207. *Eriocaulon ravenellii.*

Eriocaulon ravenelii Chapm. Map 207.

In shallow water and swamps in pine barrens. Along the coast in the southeastern states.

Eriocaulon septangulare With. Figs. 84, H–J; 85, A–B. Map 208.

Common in shallow lakes, ponds, and slow streams; mostly on sandy bottom. Northeastern states and the Great Lakes region.

Map 208. *Eriocaulon septangulare.*

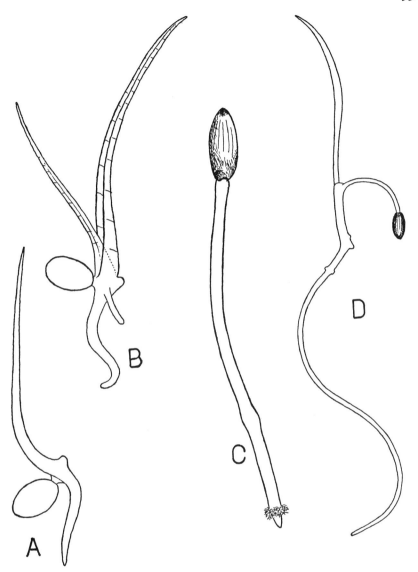

Fig. 85. *Eriocaulon septangulare* (A, B) and *Heteranthera dubia* (C, D).
(A, B) Seedlings; x 5.
(C) Germinating seed; x 10. (D) Seedling; x 3.

XYRIDACEAE: Yellow-eyed-Grass Family

Rushlike herbs with basal tufts of linear, flattened, or terete leaves sheathing the peduncle. Peduncle simple, erect, naked, usually ridged, terminating in a nearly globose head or elongate spike. Flowers perfect, solitary, and axillary in the spirally imbricated bracts. Calyx irregular; corolla regular; stamens 3, opposite the petals; 3 staminodia, when present, alternate with the stamens. Ovary 1, with 3 parietal placentae; fruit a 1-celled, 3-valved capsule; seeds many, with endosperm. — Two genera with about 50 known species.

XYRIS: Yellow-eyed Grass

Perennials or annuals with flattened, submersed or emersed, erect leaves sheathed from ⅓ to ½ their length. Flowers solitary in the axils of scalelike, imbricated bracts with narrow or broad, black or dark dorsal areas. Sepals 3, the 2 lateral keeled and persistent, the anterior larger, covering the corolla in the bud and falling with it. Petals 3, yellow or white with cohering claws. Stamens 3 fertile alternating with 3 sterile ones which are often plumose or bearded at the apex. Pistil with 1-celled ovary and 3-cleft style; capsule oblong, with 3 parietal placentae and many fusiform or oblong seeds. — About 40 species, widespread mostly in the tropics; a number of species occur in acid or sandy soils in the United States. Of these a few are aquatics or semi-aquatics.

KEY TO SPECIES OF XYRIS

1. Lateral sepals included (shorter than the subtending bract).
    2. Keel of lateral sepals ciliate; peduncle 4 to 9 dm. tall .... *X. ambigua*
    2. Keel of lateral sepals dentate, lacerate, or fimbriate.
        3. Lateral sepals with keel lacerate-fimbriate from base to apex; peduncle 6 to 10 dm. tall .... *X. iridifolia*
        3. Lateral sepals with keel lacerate-fimbriate or dentate only in the upper half; peduncles 1 to 5 dm. tall.
            4. Flowering bracts with narrow, indistinct dorsal area; plant with rhizomes; leaf tufts sparse .... *X. montana*
            4. Flowering bracts with broad, distinct dorsal area; leaf tufts dense ....
            .... *X. caroliniana*
1. Lateral sepals exserted (longer than the subtending bract); peduncles 5 to 10 dm. tall.
    5. Keel of lateral sepals slightly lacerate or erose .... *X. smalliana*
    5. Keel of lateral sepals long-lacerate or fimbriate .... *X. fimbriata*

Xyris ambigua Beyr. Fig. 86, H–I. Map 209.

On boggy shores and margins of swamps. Virginia to Texas, mostly near the coast.

Xyris caroliniana Walt. Fig. 86, A–D. Map 210.

In shallow water and on wet sandy shores of ponds and lakes. Chiefly in the northeastern states; local elsewhere.

MAP 209. *Xyris ambigua.*

MAP 210. *Xyris caroliniana.*

MAP 211. *Xyris fimbriata.*

MAP 212. *Xyris iridifolia.*

MAP 213. *Xyris montana.*

MAP 214. *Xyris smalliana.*

Xyris fimbriata Ell. Fig. 86, N–O. Map 211.

In shallow ponds and wet depressions in pine barrens. Local along the Atlantic Coast and in Texas.

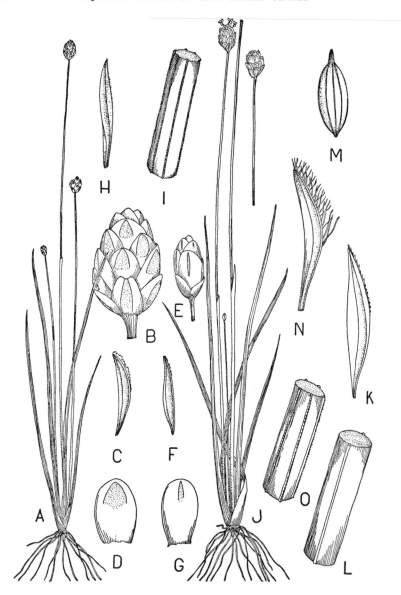

Fig. 86. *Xyris caroliniana* (A-D), *X. montana* (E-G), *X. ambigua* (H, I), *X. smalliana* (J-M), and *X. fimbriata* (N, O). (A) Plant; x 1/2. (B) Head of flowers; x 3. (C) Lateral sepal; x 5. (D) Flower bract; x 3. (E) Head; x 3. (F) Lateral sepal; x 5 (G) Flower bract; x 3. (H) Lateral sepal; x 5. (I) Lower section of peduncle; x 5. (J) Plant; x 1/2. (K) Lateral sepal; x 5. (L) Lower section of peduncle; x 5. (M) Seed; x 50. (N) Lateral sepal with fimbriate margin; x 5. (O) Lower section of peduncle; x 5.

Xyris iridifolia Chapm. Map 212.

In shallow ponds and wet depressions in pine barrens from Virginia to Texas, chiefly along the coast.

Xyris montana Ries. Fig. 86, E–G. Map 213.

In shallow water along borders of sandy ponds and in peat bogs. In the North Atlantic states.

Xyris smalliana Nash. Fig. 86, J–M. Map 214.

On soft, mucky bottom in shallow water or on floating bogs. Mostly near the coast, from Maine to Mississippi.

REFERENCE

Malme, G. O. A. Xyridaceae. In, North Amer. Flora. 19:1–15. 1937.

PONTEDERIACEAE: Pickerelweed Family

Perennials or, rarely, annuals with creeping rootstocks and fibrous roots. Leaves entire, in basal clusters or on branched, leafy stems. Flowers perfect, more or less irregular, solitary or spicate in a spathe. Perianth tubular, with a limb of 6 petal-like lobes. Stamens 3 or 6, mostly unequal, inserted on the perianth throat. Pistil with 1- to 3-celled, superior ovary and a 3- or 6-lobed stigma. Fruit a many-seeded capsule or 1-seeded and achene-like. — Four genera mostly in fresh waters of tropical America.

KEY TO GENERA

1. Leaves broad, with short, inflated, spongy petioles; plants mostly free-floating or, rarely, rooted in mud; flowers several in a spike .... *Eichhornia*
1. Leaves grasslike and sessile or broad but without inflated petioles.
   2. Plants emersed; stem simple, 1-leaved; flowers in a terminal spike; stamens 6 .... *Pontederia*
   2. Plants submersed; stem branched, leafy; flowers in axillary 1- or few-flowered spathes; stamens 3 .... *Heteranthera*

HETERANTHERA: Mud Plantain

Perennials or annuals, submersed, floating or rooted on muddy shores. Stems slender, branched, leafy, usually creeping. Leaves linear to reniform or orbicular, with sheathing petiole. Flowers solitary or a few in spathes. Perianth with a slender tube and a limb of nearly equal or unequal lobes. Stamens 3, unequal or nearly equal, inserted in the perianth throat. Capsule 1- to 3-celled, dehiscent or indehiscent, several-seeded. — About 10 species, mostly native to tropical America.

Fig. 87. *Eichhornia crassipes*.

(A) Habit sketch of a flowering plant; x 1/3. (B) Seedling protruding from seed coats. (C) Seedling. (D) Seedling with elongated hypocotyl (germinated in water). (E) Seedling without hypocotyl (germinated on soil). (B–E after Parija, 1934.)

FIG. 88. *Heteranthera dubia* (A, B), *H. reniformis* (C), and *H. limosa* (D).
(A) Plant; x 1/2. (B) Flower with spathe; x 1.
(C) Plant; x 1/2.
(D) Plant; x 1/2.

KEY TO SPECIES OF HETERANTHERA

1. Leaves grasslike, linear.
   2. Flowers yellow, nearly regular, solitary in a spathe; perianth tube 1 to 7 cm. long .... H. dubia
   2. Flowers blue and white, very irregular, several in a spathe; perianth tube 0.4 to 0.5 cm. long .... H. mexicana
1. Leaves reniform, cordate, or lanceolate; perianth blue or white.
   3. Leaves reniform to cordate; spathe 3- to 6-flowered .... H. reniformis
   3. Leaves lanceolate to nearly orbicular; spathe 1-flowered .... H. limosa

Heteranthera dubia (Jacq.) MacM. Figs. 88, A–B; 85, C–D. Map 215.

In shallow ponds, lakes, and slow rivers; chiefly in calcareous regions. Widespread in the eastern and middle western states; local in the Far West.

MAP 215. *Heteranthera dubia.*

MAP 216. *Heteranthera limosa.*

MAP 217. *Heteranthera mexicana.*

MAP 218. *Heteranthera reniformis.*

Heteranthera limosa (Sw.) Willd. Fig. 88, D. Map 216.
In ponds and lakes. Local, mostly in the Great Plains states.

Heteranthera mexicana Wats. Map 217.
In ponds and sluggish streams. Southern Texas and Mexico.

Heteranthera reniformis R. and P. Fig. 88, C. Map 218.
In shallow water, mostly on tidal mud flats of rivers and creeks. Middle Atlantic Coast, infrequent in the Mississippi Valley.

PONTEDERIA: Pickerelweed

Stout perennials with thick, creeping rootstocks with clusters of erect leaves and a simple stem with 1 leaf and a terminal spike of flowers. Leaves with fleshy, sheathing petioles and heart-shaped to lance-shaped blades with entire margin. Flowers perfect, irregular. Perianth violet-blue, funnel-form, hairy on the outside, 2-lipped; the upper lip formed by the 3 more or less fused upper lobes; the lower lip of 3 separate lobes. Stamens 6; the 3 anterior exserted; the 3 posterior, often sterile, included in the throat. Ovary 3-celled, 2 sterile, 1 fertile with a single ovule. Fruit 1-seeded, achene-like, surrounded by the perianth, with winglike crested ridges and tipped with the persistent thickened style base. — About 5 species; native to America.

Pontederia cordata L. Pickerelweed. Figs. 89; 90, C. Map 219.

Common in shallow water and on soft, mucky shores throughout the eastern states.

NOTE. *P. lanceolata* Nutt. of the southeastern states is described as having shorter fruits and a more glandular-hairy perianth. Since both of these characters are variable it seems undesirable to rank it specifically distinct from *P. cordata* (Map 220).

MAP 219. *Pontederia cordata.*

MAP 220. *Pontederia cordata* var. *lanceolata.*

EICHHORNIA: Water Hyacinth

Plants floating or rooted in mud, with slender, perennial rootstocks with rosettes of petioled leaves and fibrous, branched roots. Leaves slender and submersed or emersed and broadly ovate to cordate, with short petioles often inflated and bladder-like. Flowers several to many in showy axillary spikes or panicles, on peduncles from a 2-valved spathe; the valves very unequal, the lower broad and leaflike and the upper bractlike. Perianth showy, violet, blue, or white, tubular, with an irregular limb of 3 narrow outer lobes and 3 broad inner lobes, the

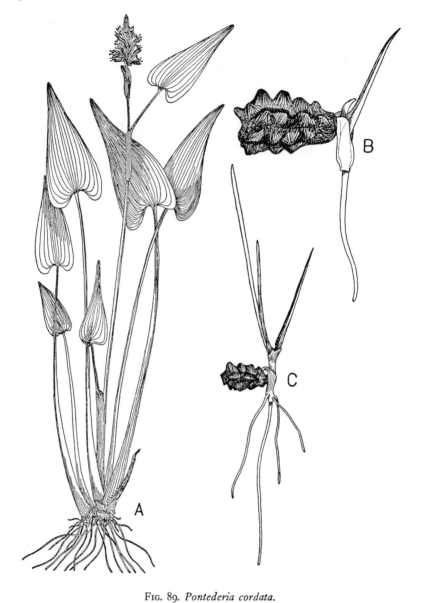

Fig. 89. *Pontederia cordata.*
(A) Plant; x 1/4. (B) Germinating seed; x 3. (C) Seedling; x 1 1/2.

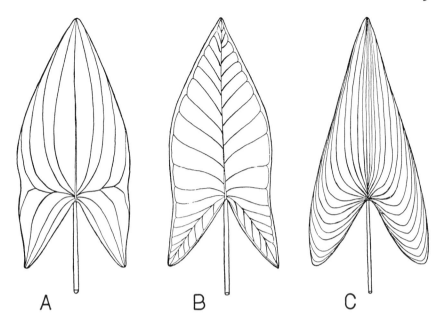

Fig. 90. Venation types in the sagittate leaves of *Sagittaria latifolia* (A), *Peltandra virginica* (B), and *Pontederia cordata* (C).

upper lobe with a yellow spot. Stamens 6; the 3 anterior with pubescent filaments, exserted, the others included in the perianth throat, with glabrous filaments. Stigma hairy; capsule 3-celled, many-seeded, 0.5 to 1.5 cm. long. — About 5 species, mostly native to South and Central America.

KEY TO SPECIES OF EICHHORNIA

1. Plants floating; flowers in spikes; petioles often inflated .... *E. crassipes*
1. Plants rooted in mud; flowers in panicles; petioles not inflated .... *E. paniculata*

Eichhornia crassipes (Mart.) Solms. Fig. 87. Map 221.

Floating in shallow ponds, lakes, and slow streams; usually considered to have been introduced, possibly also native to Florida. From Florida and along the Gulf Coast to Texas, also California; rare elsewhere. Introduced northward in summer pools and aquaria; not winter hardy.

NOTE. In Florida and in many tropical countries water hyacinth grows so luxuriantly in slow streams that it completely covers the water surface and interferes with navigation. In some navigable streams attempts are made to control the pest by mechanical or chemical methods.

Eichhornia paniculata (Spreng.) Solms. Map 222.

In shallow ponds and ditches and along muddy shores. Florida. Naturalized from cultivated plants; introduced from Brazil.

MAP 221. *Eichhornia crassipes*.

MAP 222. *Eichhornia paniculata*.

REFERENCES

Wylie, R. B. Cleistogamy in *Heteranthera dubia*. Univ. Iowa, Lab. Nat. Hist. Bull. 7:48–58. 1917.

Parija, P. Physiological investigations on water-hyacinth *Eichhornia crassipes*. Indian Jour. Agric. Sci. 4:419–428. 1934.

JUNCACEAE: RUSH FAMILY

Plants with general habit and appearance similar to grasses; leaves terete or flattened, somewhat sheathing at base, the lower bladeless and reduced to mere sheaths. Flowers in cymose clusters, terminal or appearing lateral, then the upper leaf erect and appearing as a continuation of the scape. Flowers perfect, regular; perianth of 3 small, glumelike, persistent sepals and 3 similar petals; stamens 6 or sometimes only 3. Ovary superior, 3-celled or 1-celled with 3 parietal placentae; style short, tipped with 3 filiform, hairy stigmas; fruit a 3-valved, loculicidal capsule with many or only 3 seeds. — Eight genera with many widely distributed species.

JUNCUS: RUSH

Mostly glabrous perennials with creeping rootstocks, rarely annuals. Stems mostly simple, pithy, hollow, or with partitions. Leaves often hollow or nodose, spearlike, sometimes flat. Flowers small, green or brownish; stamens 6 or 3, capsule 3-celled or imperfectly so, many-seeded. Seeds with or without caudate appendages. — About 200 species, most common in the Northern Hemisphere; many occur in marshes and bogs, a few in water.

KEY TO SPECIES OF JUNCUS *Good only when stamens available as well as seeds!*

1. Cyme appearing lateral; leaves reduced to sheaths; involucral bract appearing as continuation of stem .... *J. balticus*
1. Cyme or flower clusters terminal; at least some of the leaves with blades.
    2. Annuals; leaves terete, nonseptate; stem creeping or often floating ...*J. repens*
    2. Perennials.
        3. Leaves flat or, if terete, nonseptate.
            4. Leaf sheath covering ½ the stem or more; flowers with 2 bracteoles in addition to the bractlet at base of pedicel; sepals obtuse .... *J. gerardi*
            4. Leaf sheath covering ⅓ the stem or less; flowers with 1 bractlet at base of pedicel; sepals acute or acuminate ....*J. marginatus*
        3. Leaves hollow or septate, terete or flattened.
            5. Stamens 3.
                6. Seeds caudate .... *J. canadensis*
                6. Seeds not caudate.
                    7. Leaves flattened .... *J. polycephalus*
                    7. Leaves terete .... *J. acuminatus*
            5. Stamens 6.
                8. Upper cauline leaves bladeless, of firm, tawny sheaths; middle leaf erect, longer than inflorescence .... *J. militaris*
                8. Upper cauline leaves with blades; flowers solitary or in 2's, often replaced by leafy buds.
                    9. Stems erect from creeping rootstocks .... *J. pelocarpus*
                    9. Stems lax, floating, with tufts of roots at nodes .... *J. subtilis*

Juncus acuminatus Michx. Fig. 91, A–C. Map 223.

Common in marshes, sloughs, and shallow water of ponds and ditches. Widespread, mostly in the eastern states.

MAP 223. *Juncus acuminatus.*   MAP 224. *Juncus balticus.*

Juncus balticus Michx. Fig. 91, G–I. Map 224.

In marshes and swales and along streams. Across the northern United States.

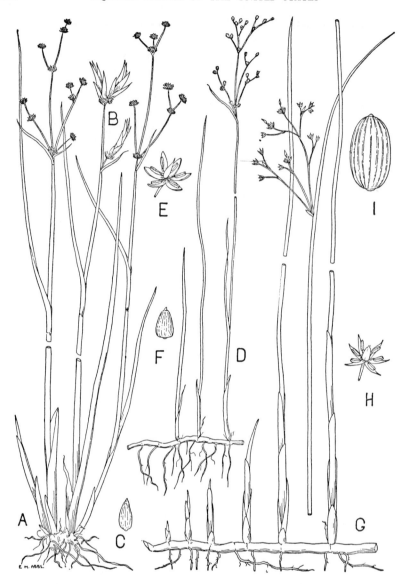

FIG. 91. *Juncus acuminatus* (A–C), *J. gerardi* (D–F), and *J. balticus* (G–I).
(A) Plant; x 1/2. (B) Vegetative buds; x 1/2. (C) Seed; x 20.
(D) Plant; x 1/2. (E) Flower; x 3. (F) Seed; x 20.
(G) Plant with rhizome; x 1/2. (H) Flower; x 3. (I) Seed; x 20.

Fig. 92. *Juncus militaris* (A–C), *J. marginatus* (D, E), and *J. polycephalus* (F). (A) Plant with rhizome; x 1/2. (B) Flower; x 3. (C) Seed; x 20. (C') Seedling; x 7. (D) Plant with rhizome; x 1/2. (E) Seed; x 20. (F) Plant; x 1/2.

Juncus canadensis Gay. Fig. 93, F–G. Map 225.

Common in marshes, sloughs, and shallow water. In the Atlantic Coast states and the Mississippi Valley.

Juncus gerardi Lois. Black grass. Fig. 91, D–F. Map 226.

Mostly in salt marshes and on brackish shores. Atlantic Coast; local inland and along the Pacific Coast.

Juncus marginatus Rostk. Fig. 92, D–E. Map 227.

Widespread in ponds, streams, and marshes; throughout the eastern states.

Juncus militaris Bigel. Bayonet rush. Fig. 92, A–C. Map 228.

In shallow ponds, mostly on sandy bottom; often forming extensive meadows of sterile plants when growing in water about 1 meter deep. North Atlantic states; local about the Great Lakes.

Juncus pelocarpus Meyer. Fig. 93, A–C. Map 229.

In shallow ponds and along sandy shores. Northeastern states and about the Great Lakes.

NOTE. In New York and New England many of the clear lakes with sandy bottom and acid water support extensive beds of this species extending from the shore to depths of 1 to 4 meters. In such localities the plants are sterile and form a sodlike mat only about 5 to 10 cm. high. These plants increase and spread by slender, creeping rootstocks. Plants growing emersed in shallow water or on exposed shores from which the water has receded flower and set seed in abundance. Such plants sometimes also produce numerous small leaf buds in the axils of the inflorescence. After falling, these buds may be washed away and later take root on a bottom or a moist shore upon which they have come to rest.

Juncus polycephalus M.A.Curtis. Fig. 92, F. Map 230.

In swamps and along borders of streams. Chiefly in the southeastern states and westward to Texas.

Juncus repens Michx. Fig. 93, D–E. Map 231.

In shallow ponds and along streams. South Atlantic and Gulf Coast states.

Juncus subtilis Meyer. Map 232.

In shallow water and on margins of ponds and streams. Maine and northward.

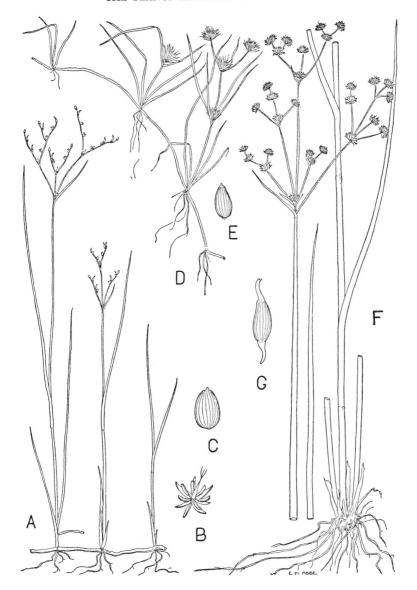

Fig. 93. *Juncus pelocarpus* (A–C), *J. repens* (D, E), and *J. canadensis* (F, G).
(A) Plant with rootstock; x 1/2. (B) Flower; x 3. (C) Seed; x 20.
(D) Plant; x 1/2. (E) Seed; x 20.
(F) Plant; x 1/2. (G) Seed with caudate appendages; x 20.

Map 225. *Juncus canadensis.*

Map 226. *Juncus gerardi.*

Map 227. *Juncus marginatus.*

Map 228. *Juncus militaris.*

Map 229. *Juncus pelocarpus.*

Map 230. *Juncus polycephalus.*

Map 231. *Juncus repens.*

Map 232. *Juncus subtilis.*

MARANTACEAE: Arrowroot Family

Coarse herbs with large, 2-ranked, pinnately veined leaves. Flowers in panicles, perfect and irregular. Sepals 3, nearly separate; petals 3, nearly separate; stamens reduced, only ½ of 1 anther bearing pollen, the other anthers sterile, petaloid. Compound pistil with inferior, 3-celled ovary; each cell with 1 ovule. Style solitary, more or less curved or 1-sided. Seeds with aril and large endosperm. — About 10 genera, mostly of tropical regions.

THALIA

Leaves basal, with long, erect petiole and ovate-lanceolate blades with entire margin. Scape about 1 to 2 meters high, terminated by a panicle with colored bracts. Sepals nearly equal; petals blue, longer than the sepals. Staminodia more or less fused, petaloid, unequal, purple or violet. Fruit subglobose to ovoid. — Six species, American.

Map 233. *Thalia dealbata.*

Thalia dealbata Roscoe. Map 233.

In shallow ponds, sloughs, and swamps. Mostly on the Coastal Plain from Florida to Texas; northward to Missouri, also South Carolina.

POLYGONACEAE: Buckwheat Family

Plants with jointed stems and alternate, simple, entire leaves with stipules fused into sheaths (ocreae). Flowers mostly perfect, regular; calyx 3- to 6-cleft or parted; stamens 4 to 12; pistil with 1-celled, superior ovary and 2 or 3 styles. Fruit a hard achene, usually 3-angled, lenticular or plano-convex. — A large family, mostly of terrestrial members; aquatic members represented in the United States by only a few species of the following genus.

POLYGONUM: Smartweed, Knotweed

Perennials or annuals with stems swollen at the nodes, erect, or with creeping rootstocks. Nodes covered by the thin, cylindrical, stipular sheaths. Flowers in open or crowded spikes with small bracts; calyx pink, white, or greenish, 4- to 5-parted, appressed to the mature achene; stamens 4 to 8. Achene glossy or dull, mostly dark brown or black, with endosperm. — About 250 species, many widely distributed, only a few aquatic.

KEY TO SPECIES OF POLYGONUM

1. Sepals glandular-dotted; flowers greenish-white, in slender, open spikes .... *J. punctatum*
1. Sepals not glandular-dotted.
   2. Spikes several on a peduncle .... *P. densiflorum*
   2. Spikes solitary or 2 on a peduncle; flowers pink or pure white.
      3. Leaves narrowly lanceolate; spikes slender and open; sheaths mostly ciliate along the margin .... *P. hydropiperoides*
      3. Leaves elliptical to ovate-lanceolate; spikes of densely crowded flowers; sheaths often not ciliate.
         4. Leaves elliptical to oblong-lanceolate, with obtuse apex; spike 1 to 2.5 cm. long .... *P. amphibium*
         4. Leaves ovate-oblong to ovate-lanceolate, with acute or acuminate apex; spike 3 to 10 cm. long .... *P. coccineum*

Polygonum amphibium L. Water smartweed. Fig. 94, A–C. Map 234.

Common in shallow ponds, lakes, and reservoirs, usually on soft bottom; throughout the northern states.

NOTE. This species usually grows rooted on the bottom, but in reservoirs and lakes with greatly fluctuating water level it may grow as floating mats of intertangled rootstocks, rising and falling with the water. Aquatic or floating forms have glabrous and glossy leaves. Terrestrial forms are more erect and have leaves and stems hairy and a sheath with a foliaceous border, forma *terrestris*, or a sheath without a border, forma *hartwrightii*.

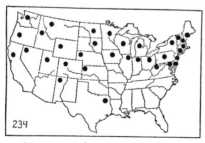

MAP 234. *Polygonum amphibium*.　　MAP 235. *Polygonum coccineum*.

Polygonum coccineum Muhl. Fig. 94, D-F. Map 235.

Locally common in shallow ponds, sluggish streams, marshes, and wet muck lands. Throughout all but the southern states.

NOTE. This species is very similar to the preceding in habit. It also has a hairy terrestrial state, forma *terrestris*, which occurs more commonly than the glabrous and glossy aquatic state, forma *natans*. At times both states may be observed growing out of the branches of the same rootstock, one growing on shore, the other in the water.

MAP 236. *Polygonum densiflorum*. MAP 237. *Polygonum hydropiperoides*.

Polygonum densiflorum Meisn. (*P. portoricensis* [Bertero] Small). Map 236.

In ponds, marshes, and wet shores, chiefly on the Coastal Plain. Florida to Texas and northward to Missouri and New Jersey.

Polygonum hydropiperoides Michx. Water pepper. Fig. 95, D-F. Map 237.

In spring-fed ponds and bog pools and on marshy shores of streams. Throughout the area east of the Great Plains; local on the Pacific Coast.

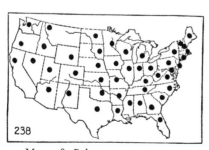

MAP 238. *Polygonum punctatum*.

Polygonum punctatum Ell. Fig. 95, A-C. Map 238.

In marshy places and sloughs and on wet shores. Throughout the United States.

Fig. 94. *Polygonum amphibium* (A–C) and *P. coccineum* (D–F).

(A) Plant; x 1/2. (B) Flower; x 10. (C) Achene; x 10.
(D) Plant; x 1/2. (E) Achene; x 10. (F) Leaf showing fused sheathing stipules (ocrea); x 1/2.

FIG. 95. *Polygonum punctatum* (A–C) and *P. hydropiperoides* (D–F). (A) Plant; x 1/2. (B) Flowers with punctate sepals; x 5. (C) Achene; x 10. (D) Plant; x 1/2. (E) Flower; x 5. (F) Achene; x 10.

REFERENCE

Stanford, E. E. The amphibious group of *Polygonum*, subgenus *Persicaria*. Rhodora 27:109–112, 125–130, 146–152, 156–166. 1925.

CHENOPODIACEAE: GOOSEFOOT FAMILY

Plants more or less succulent, leaves alternate or opposite, fleshy or reduced to thickened scales, without stipules. Flowers small, greenish; calyx mostly 3- to 5-cleft or parted; stamens usually the same number as calyx parts and opposite them. Pistil with 1-celled ovary, forming a utricle. Seed with embryo coiled or spiral; endosperm central or wanting. — A large group of terrestrial plants; a few species occur in salt marshes or tidal waters.

KEY TO GENERA

1. Leaves opposite, fleshy, reduced to scales or teeth .... *Salicornia*
1. Leaves alternate, succulent, linear, terete or nearly so .... *Suaeda*

SALICORNIA: GLASSWORT, SAMPHIRE

Low annuals or perennials with succulent, branched stems and reduced, scalelike leaves. Flowers reduced and sunken in the upper axils on the fleshy axis which forms a thick spike, mostly 3 in each axil, the central perfect, the 2 lateral sometimes staminate. Calyx utricle-like, persisting. Stamens 1 or 2; pistil 1-celled, with 2 styles; fruit 1-seeded, pubescent. Seed erect, without endosperm. — About 10 species of saline plants.

KEY TO SPECIES OF SALICORNIA

1. Plants perennial; all flowers in the axil of a scale nearly equal in height.
    2. Spikes 4 mm. thick, not more than 2 cm. long .... *S. utahensis*
    2. Spikes slender, less than 4 mm. thick, usually more than 2 cm. long *S. perennis*
1. Plants annual; the middle flower in the axil of a scale higher than the 2 lateral flowers.
    3. Scales mucronate or sharp-pointed .... *S. bigelovii*
    3. Scales rounded or blunt at apex.
        4. The pair of opposite scales of each joint wider than high; stem grayish-green .... *S. europaea*
        4. The pair of opposite scales of each joint about as wide as high; stem reddish .... *S. rubra*

Salicornia bigelovii Torr. Fig. 96, E. Map 239.

In salt marshes and along brackish shores. Along the Atlantic and Gulf Coasts; local in Arizona and California.

Salicornia europaea L. Fig. 96, A–D. Map 240.

In salt marshes and tidal flats. Along the Atlantic Coast; rare on salt flats inland.

Salicornia perennis Mill. Fig. 97, A–B. Map 241.

In salt marshes and meadows and on beaches. Common on the Atlantic and Pacific Coasts.

Salicornia rubra A.Nels. Fig. 96, F. Map 242.

On salt flats and borders of alkaline ponds and lakes. Chiefly in the Rocky Mountain region, infrequent on the Great Plains.

Salicornia utahensis Tidestrom. Fig. 97, C–D. Map 241.

Local along shores of salt lakes and on salt flats and marshes in the Great Basin.

MAP 239. *Salicornia bigelovii.*

MAP 240. *Salicornia europaea.*

MAP 241. *Salicornia perennis,* ●; S. *utahensis,* ▲.

MAP 242. *Salicornia rubra.*

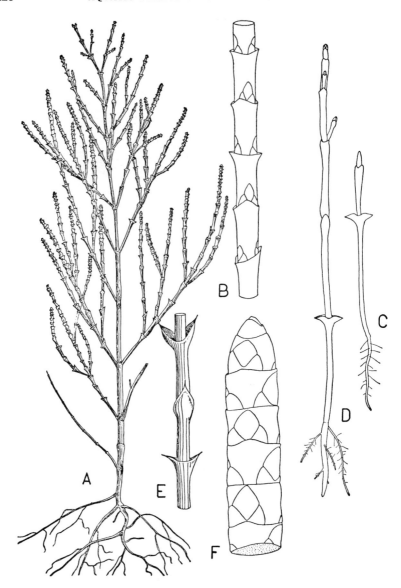

Fig. 96. *Salicornia europaea* (A–D), *S. bigelovii* (E), and *S. rubra* (F).
(A) Plant; x 1/2. (B) Section of flowering spike; x 5. (C, D) Seedlings; x 1.
(E) Section of stem; x 2.
(F) Section of flowering spike; x 5.

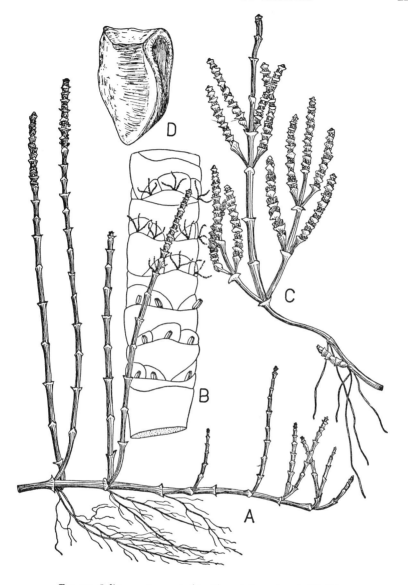

Fig. 97. *Salicornia perennis* (A, B) and *S. utahensis* (C, D).
(A) Plant; x 1. (B) Flower spike with pistillate flowers above and staminate flowers below; x 5.
(C) Plant; x 1. (D) "Seed" (utricle); x 15.

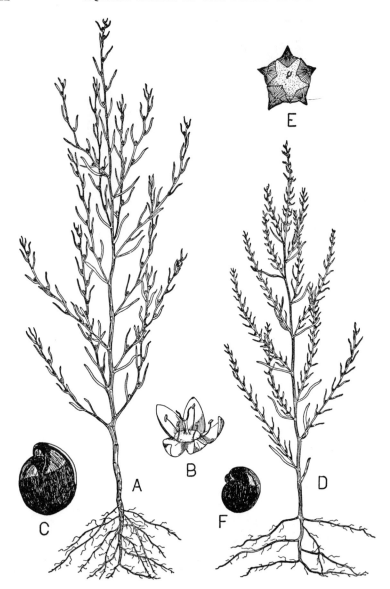

Fig. 98. *Suaeda linearis* (A–C) and *S. maritima* (D–F).
(A) Plant; x 1/3. (B) Flower; x 6. (C) Seed; x 10.
(D) Plant; x 1/3. (E) Flower; x 5. (F) Seed; x 10.

SUAEDA (*DONDIA*): Sea Blite

Plants with alternate, fleshy, linear, nearly terete leaves. Flowers axillary or in axillary clusters, small, greenish. Calyx 5-parted, fleshy, often keeled or crested; stamens 5; pistil with 2 or 3 stigmas. Seed usually black and glossy, lenticular, with a flat, spiral embryo; endosperm scanty or absent. — Plants mostly of alkaline soils or salt marshes and meadows.

KEY TO SPECIES OF SUAEDA

1. Seed 2 mm. in diameter; plant mostly erect and with branches nearly simple; leaves mostly glaucous .... S. *maritima*
1. Seed 1 to 1.5 mm. in diameter; plant mostly spreading, or ascending, with branches much-divided; leaves not glaucous .... S. *linearis*

Suaeda linearis (Ell.) Moq. Fig. 98, A–C. Map 244.

In salt marshes and on sandy beaches. Along the Atlantic and Gulf Coast.

Suaeda maritima (L.) Dumort. Fig. 98, D–F. Map 243.

In salt marshes, brackish bays, and tidal mud flats. Along the coast from New England to New Jersey. Perhaps naturalized from Europe.

Map 243. *Suaeda maritima.*

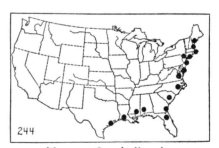

Map 244. *Suaeda linearis.*

REFERENCE

Fernald, M. L. The genus *Suaeda* in northeastern America. Rhodora 9:140–146. 1907.

AMARANTHACEAE: Amaranth Family

Herbs with alternate or opposite, simple leaves without stipules. Flowers and fruits similar in structure to those of Chenopodiaceae but subtended by dry scarious and persistent bracts. — A large family best represented in tropical regions; a few species occur in marshes and water.

KEY TO GENERA

1. Plant creeping or forming floating mats; leaves opposite; flowers in axillary spikes .... *Alternanthera*
1. Plant erect; leaves alternate; flowers in panicles .... *Acnida*

ACNIDA: WATER HEMP

Coarse, erect, dioecious annuals, from 1 to 2 meters high, with alternate, entire leaves. Flowers in large terminal panicles, 3-bracted, those of the staminate plants with calyx of 5 thin sepals exceeding the bracts; stamens 5. Pistillate flowers naked; pistil with superior ovary and 2 to 5 long, plumose stigmas. Fruit coriaceous or a thin-walled, 3- to 5-angled utricle, dehiscent or indehiscent.

Acnida cannabina L. Fig. 100, A–D. Map 246.

In salt marshes and tidal bays and on banks of streams. Chiefly near the coast, from Maine to Florida.

MAP 245. *Alternanthera philoxeroides*.   MAP 246. *Acnida cannabina*.

ALTERNANTHERA

Stems prostrate or creeping, branched, and often jointed and rooting at the nodes. Leaves opposite, mostly entire, linear to oblong-lanceolate. Flowers in sessile or peduncled spikes, perfect. Perianth of 5 unequal, separate or nearly separate, pale green or whitish sepals. Stamens 5, their filaments fused at the base. Fruit a utricle. — Several species, mostly of dry places in tropical regions.

Alternanthera philoxeroides (Mart.) Griseb. (*Achyranthes philoxeroides* [Mart.] Standl.). Alligator weed. Fig. 99. Map 245.

Locally abundant in ponds, ditches, bays, and sluggish streams. Southeastern states.

THE PLANTS ARRANGED BY FAMILIES 225

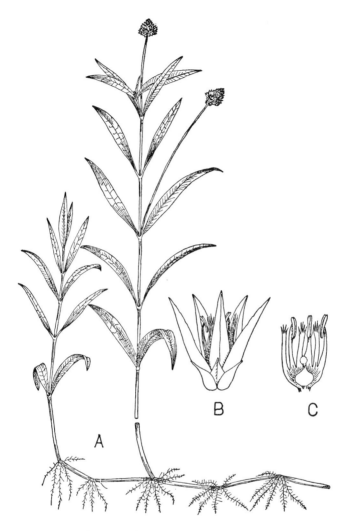

Fig. 99. *Alternanthera philoxeroides.*

(A) Habit sketch of small plant; x 1/4. (B) Flower; x 3.
(C) Vertical section of flower with perianth removed; x 3.

NOTE. A spreading weed, sometimes forming floating mats over extensive areas crowding out other plants. During low water it covers muddy banks and lowlands. Introduced northward but not hardy in the northern states.

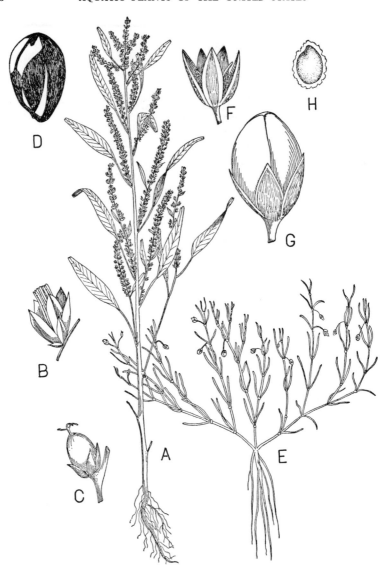

Fig. 100. *Acnida cannabina* (A–D) and *Spergularia canadensis* var. *occidentalis* (E–H).
(A) Plant; x 1/4. (B) Staminate flower; x 5. (C) Pistillate flower; x 5. (D) Seed; x 10.
(E) Plant; x 1/2. (F) Flower; x 5. (G) Capsule; x 5. (H) Winged seed; x 10.

CARYOPHYLLACEAE: Pink Family

Herbs with opposite, simple, entire leaves, without or rarely with stipules; stems mostly with swollen nodes. Flowers in cymes or axillary, regular, perfect or, rarely, imperfect. Calyx of 4 or 5 separate or fused persistent sepals; corolla of 5 petals or, rarely, absent; stamens usually 5 or 10, separate; pistil with 2 to 5 styles and a 1-celled, superior ovary with free-central placenta. Fruit a capsule with many seeds.

SPERGULARIA: Sand Spurry

Low, spreading annuals or perennials with linear, often somewhat fleshy, terete leaves with scaly stipules. Flowers small, axillary; sepals 5; petals 5, entire, pink or white; stamens 2 to 10; pistil with 3 styles, forming a short, 3-valved, many-seeded capsule. Seeds flattened, often wing-margined. — Several species, mostly restricted to salt marshes and sandy seabeaches.

Spergularia canadensis (Pers.) G. Don. Map 247.

In salt marshes and on brackish tidal shores; New England Coast to Long Island, New York. *S. canadensis* var. *occidentalis* Rossbach, with more erect habit, glandular, pubescent stems, and large stipules, occurs

Map 247. *Spergularia canadensis*, East;
S. canadensis var. *occidentalis*, West.

along the Pacific Coast in open depressions in salt marshes and also on tidal mud flats where it may be completely submersed most of the time (Fig. 100, E–H).

REFERENCE

Rossbach, R. P. *Spergularia* in North and South America. Rhodora 42:57–83, 105–143, 158–193, 203–213. 1940.

CERATOPHYLLACEAE: Hornwort Family

Submersed, rootless, olive-green to green herbs with a slender main axis from 1 to 20 dm. long and scattered lateral branches. The lower end of the stem is frequently imbedded in soft mud and usually without chlorophyll. Leaves in whorls, divided into slender, often stiff and brittle segments, much crowded toward the apex by the shortening of the internodes, giving the shoot the appearance of a "coontail." Monoecious plants with reduced staminate and pistillate inflorescences on different nodes, rarely on the same node. The inflorescences are not in the axils of leaves but take the place of a leaf in the whorl. — The family consists of only one genus.

CERATOPHYLLUM: Hornwort, Coontail

Staminate inflorescence of numerous (8 to 18) distinct stamens crowded in whorls on a short, convex receptacle, surrounded by an involucre of 10 to 12 elongate, forked, persistent bracts. Staminate flower consisting of a naked stamen with short filament and oblong-linear, bilocular anther with longitudinal dehiscence. The filament ends in a connective with terminal expansion which acts as a float when the mature anthers become detached and rise to the water surface. After dehiscence of the anther the pollen sinks and comes in contact with the submersed pistillate flower. Pistillate inflorescence of a solitary, naked pistil surrounded by an involucre of 10 to 20 bristle-pointed bracts. Occasionally an involucre may contain, in addition to the normally developed pistil, 1 or more reduced lateral pistils each with its own whorl of bracts indicating a much-reduced compound inflorescence. Pistil with slender, persistent style and 1-celled ovary. Fruit a laterally compressed achene, 5 to 7 mm. long, with persistent involucre, a terminal spine, and 2 basal spines, also marginal lateral spines in some species. Seed with a well-developed embryo consisting of a pair of thick cotyledons and plumule with several nodes of leaves. — Two distinct species occur in the United States. Both are widespread and common in ponds, shallow lakes, and sluggish streams. They are among the dominant aquatics in temporary ponds and newly made lakes in which the water is rich in organic matter and materials in solution.

NOTE. Early in the growing season the plants grow upright with the lower part of the stem serving for anchor and probably also for the absorption of nutrients. Later in the season many of the plants grow more or less floating near the surface of the water intertangled with filamentous algae. The tips of the branches may become much shortened and thickened, break off, and sink to the bottom to act as "winter-buds" and develop into new plants. In deep water vegetative plants may be found throughout the winter even under ice. The roots are

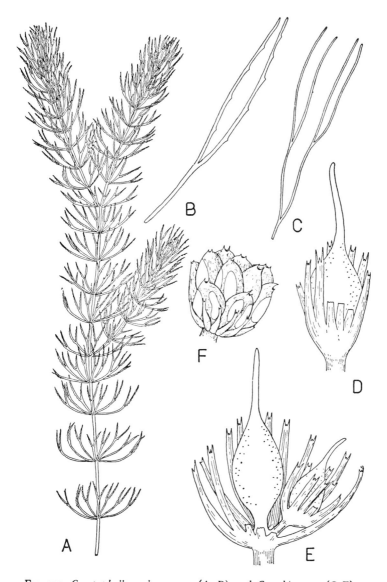

FIG. 101. *Ceratophyllum demersum* (A, B) and *C. echinatum* (C-F).

(A) Sterile branch with whorled leaves; x 1. (B) Leaf; x 5.
(C) Leaf; x 5. (D) Pistillate inflorescence, reduced to a single pistil surrounded by involucre; x 12. (E) Inflorescence with two pistillate flowers; x 12. (F) Staminate inflorescence; x 10. (D-F from Aboy, 1936.)

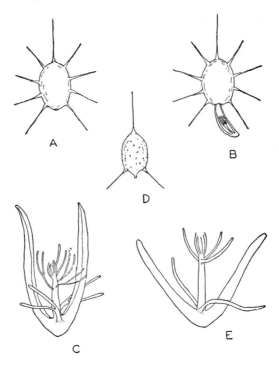

Fig. 102. Fruits and seedlings of *Ceratophyllum echinatum* (A–C) and *C. demersum* (D, E).

(A) Mature achene; (B) Seed in process of germination. (C) Seedling showing 2 fleshy cotyledons and plumule (note absence of hypocotyl); about x 2.

(D) Mature achene. (E) Seedling with simple leaves; about x 2.

(From Muenscher, 1940.)

absent even in the seedling. The radicle does not enlarge or elongate during seed germination (Fig. 102). The achenes have a hard, durable covering. They are often found in muck soils in the bottoms of former ponds and lakes. The achenes are eaten by wild ducks; large numbers have been recovered from the birds' gizzards.

KEY TO SPECIES OF CERATOPHYLLUM

1. Leaf segments subcapillary, mostly entire, often inflated; achene with 3 to 5 lateral spines on each side; seedling with leaves of all nodes divided *C. echinatum*
1. Leaf segments capillary to linear and flattened, serrate; achene without lateral spines; seedling with leaves of the lower nodes simple, linear ...... *C. demersum*

Ceratophyllum demersum L. Figs. 101, A–B; 102, D–E; 102a, A. Map 248.

In shallow ponds and slow streams. Throughout the United States.

Ceratophyllum echinatum Gray. Figs. 101, C–F; 102, A–C. Map 249.
In shallow ponds, sluggish streams, ditches, and temporary pools. Chiefly in the eastern United States.

Map 248. *Ceratophyllum demersum*.

Map 249. *Ceratophyllum echinatum*.

REFERENCES

Aboy, H. E. A study of the anatomy and morphology of *Ceratophyllum demersum*. Cornell Univ. Thesis. 1936.
Muenscher, W. C. Fruits and seedlings of *Ceratophyllum*. Amer. Jour. Bot. 27:231–233. 2 figs. 1940.
Fernald, M. L. *Ceratophyllum*. Rhodora 43:551–552. 1941.

NYMPHAEACEAE: WATER-LILY FAMILY

Perennial aquatic herbs with creeping or erect, often stout, branching rootstocks; leaves mostly alternate, cordate or peltate, sometimes very large and emersed but mostly floating or, rarely, submersed. Flowers axillary, solitary, usually large and showy; sepals and petals from 3 to numerous; stamens from 3 or 4 to many; carpels from 2 or 3 to many, free or fused into a compound pistil; fruit a many-seeded berry or 1- or 2-seeded and nutlike.

KEY TO GENERA

1. Leaves mostly dissected into linear segments .... *Cabomba*
1. Leaves all simple, entire or nearly so.
   2. Leaves peltate; carpels several, free from each other, with 1 or 2 seeds.
      3. Blade elliptic, 1 dm. or less in length; stem submersed .... *Brasenia*
      3. Blade orbicular, 2 to 6 dm. in diameter; rootstock creeping .... *Nelumbo*
   2. Leaves not peltate; carpels several, fused into a compound pistil with many seeds.
      4. Veins mostly radiating from the base of the blade, repeatedly forked; stamens and petals attached to the side and near the summit of the ovary .... *Nymphaea*
      4. Veins mostly attached to the midrib of the blade, forking but little; petals and stamens attached to the receptacle below the ovary .... *Nuphar*

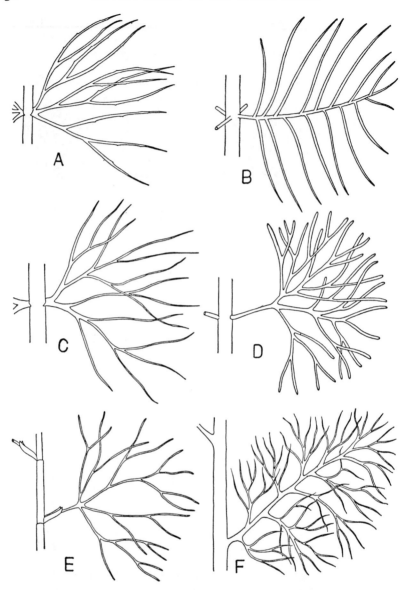

Fig. 102a. Diagrams showing differences in the dissected leaves with filiform segments of several species. (A) *Ceratophyllum demersum*. (B) *Myriophyllum exalbescens*. (C) *Bidens beckii*. (D) *Cabomba caroliniana*. (E) *Ranunculus aquatilis*. (F) *Rorippa aquatica*.

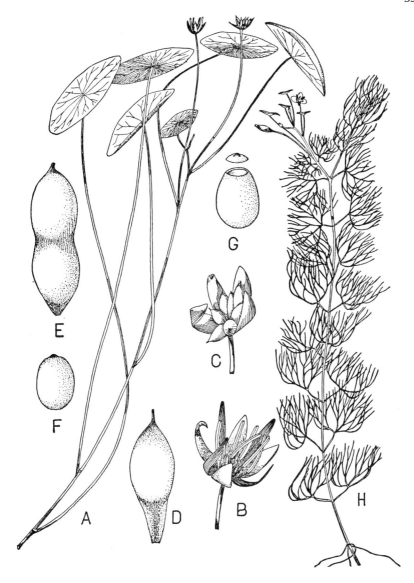

Fig. 103. *Brasenia schreberi* (A–G) and *Cabomba caroliniana* (H).

(A) Upper part of a plant; x 1/2. (B) Flower; x 1. (C) Cluster of separate carpels; x 1. (D, E) Fruits with 1 and 2 seeds; x 5. (F) Seed; x 5. (G) Seed with circumscissile lid, from peat on lake bottom; x 5.

(H) Part of a flowering plant showing 2 kinds of leaves; x 1/2.

BRASENIA: WATER SHIELD

Plants with slender, creeping rootstocks bearing slender, branched, leafy stems; leaves alternate; blade 2 to 10 cm. long, peltate, oval to elliptic, entire or, rarely, crenate, floating, mucilaginous on the undersurface. Flowers axillary, about 2 cm. in diameter; sepals 3 or 4; petals 3 or 4, linear, purplish; stamens 12 to 18; carpels 4 to 20, free, each forming a 1- or 2-seeded nutlet ultimately breaking open by a lid.

Brasenia schreberi Gmel. Water shield. Fig. 103, A–G. Map 250.

Shallow lakes, ponds, and sluggish streams; most abundant in ponds and lakes with acid water and bottom of sand or partially decomposed plant remains. Chiefly in the eastern states.

CABOMBA

Delicate herbs with slender, branched stems with opposite or whorled, finely dissected, submersed leaves and occasionally the upper leaves alternate, with entire, linear-oblong, peltate, floating blades. Flowers small, white or yellowish, solitary on slender, axillary peduncles; sepals and petals 3; stamens 3 to 6; carpels 2 to 4, free; fruits 3-seeded, indehiscent.

Cabomba caroliniana Gray. Fanwort. Figs. 103, H; 102a, D. Map 251.

In ponds and quiet bays of lakes. Chiefly in the southeastern states. This is a common aquarium plant and is the "fish grass" of the trade. It is frequently used in fish bowls as an oxygenating plant.

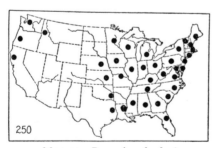
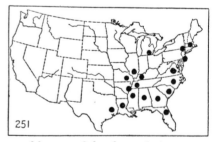

MAP 250. *Brasenia schreberi.*   MAP 251. *Cabomba caroliniana.*

NELUMBO: LOTUS

Stout herbs with long, slender, branched rootstocks with tubers and alternate leaves; the blade often emersed high above the water surface, orbicular, peltate, 2 to 9 dm. in diameter, with margin usually turned up to make a saucer-shaped depression, prominently veined from the

THE PLANTS ARRANGED BY FAMILIES 235

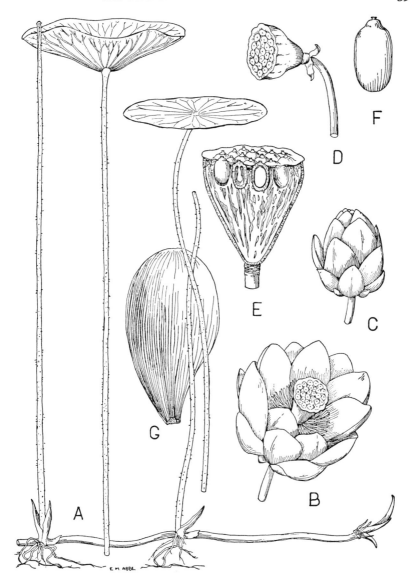

Fig. 104. *Nelumbo pentapetala.*

(A) Part of a plant with creeping rootstock; x 1/6. (B) Flower; x 1/4. (C) Flower bud; x 1/4. (D) Receptacle with imbedded carpels; x 1/4. (E) Vertical section of receptacle; x 1/2. (F) "Seed"; x 1. (G) Petal; x 1/2.

center. Flowers 1 to 3 dm. in diameter, pale yellow; sepals, petals, and stamens numerous, inserted on the receptacle below the carpels. Carpels several, surrounded by and imbedded in cavities of the large, obconical, fleshy receptacle which when mature is from 3 to 10 cm. in diameter, top-shaped, and bears the several nutlike fruits about 2 cm. long with a single seed with large embryo but no endosperm.

Nelumbo pentapetala Walt. (*N. lutea* [Willd.] Pers.). American lotus. Fig. 104. Map 252.

In shallow bays and on muddy shores in coves of sluggish streams; widespread in the eastern United States. Some of the localities where

MAP 252. *Nelumbo pentapetala.*

it is now found probably represent stations where it was introduced by the Indians who used both the seeds and tubers for food.

NUPHAR (*NYMPHAEA* L.; *NYMPHOZANTHUS* L. C. Richard): YELLOW WATER LILY, SPATTERDOCK

Perennial herbs with stout, creeping rootstocks with scattered roots and alternate leaves; the blade entire, with a deep basal sinus, orbicular, ovate, or nearly lanceolate, pinnately veined, floating or emersed, on a slender or stout, cylindric or flattened petiole. Submersed leaves with short petioles and delicate, thin, filmy blades, common on seedling plants and basal on larger plants. Flowers large, solitary on a thick peduncle; sepals 5 to 12, yellow at least on the inside; petals linear to oblong, stamen-like or scalelike, numerous, inserted on the receptacle under the ovary. Pistil several-celled, with a short style and disklike stigma with 5 to many rays; fruit ovoid to columnar, usually ripening above the water surface; seeds numerous, borne in thin, baglike sections of the ovary, mostly ovoid, yellow or brown, glossy. — About 25 species some of which are not very distinct.

KEY TO SPECIES OF NUPHAR

1. Blade of leaf less than ½ as wide as long .... *N. sagittifolium*
1. Blade of leaf more than ½ as wide as long.
    2. Sepals 9 (rarely 7) .... *N. polysepalum*
    2. Sepals mostly 6.
        3. Petioles terete or nearly so; leaves erect or floating .... *N. advena*
        3. Petioles flattened; leaves usually floating.
            4. Anthers at least as long as the filaments; flowers 4 to 5 cm. in diameter
            .... *N. variegatum*
            4. Anthers shorter than the filaments; flowers 1.5 to 3.5 cm. in diameter.
                5. Flower less than 2 cm. in diameter; stigma rays 10 or fewer ....
                .... *N. microphyllum*
                5. Flower 2.5 to 3.5 cm. in diameter; stigma rays more than 10 ....
                .... *N. rubrodiscum*

Nuphar advena Ait. Figs. 105, A–F; 110, C–E. Map 253.

Common at lower elevation in muddy bays and ponds; also found in backwaters and on tidal flats of larger streams. Widespread from the Mississippi River to the Atlantic Coast. This is a variable species, and a number of segregates, based largely upon leaf shape and habit, have been named as new species by Miller and Standley.

Nuphar microphyllum (Pers.) Fernald. Fig. 106, A–D. Map 254.

Along deeper water channels where streams enter lakes, infrequent in ponds. Northeastern states and infrequent about the Great Lakes.

Map 253. *Nuphar advena.*

Map 254. *Nuphar microphyllum.*

Nuphar polysepalum Engelm. Fig. 107, C–E. Map 255.

In ponds, shallow bays of lakes, sluggish streams, and backwaters. From the northern Rocky Mountains to the Pacific Coast.

Nuphar rubrodiscum Morong. Fig. 106, E–H. Map 256.

In shallow ponds and streams. Northeastern states and local about the Great Lakes.

238 AQUATIC PLANTS OF THE UNITED STATES

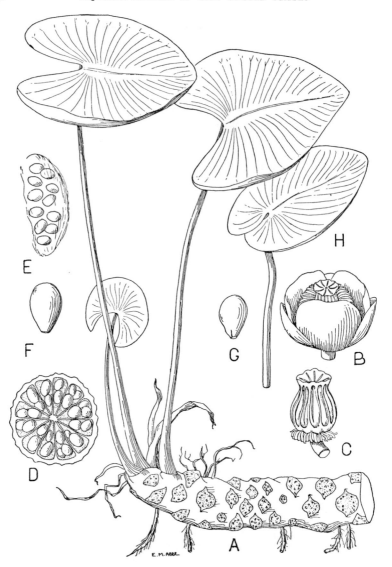

Fig. 105. *Nuphar advena* (A–F) and *N. variegatum* (G, H).

(A) Plant with thick rootstock; x 1/4. (B) Flower; x 1/2. (C) Fruit; x 1/2. (D) Cross section of fruit; x 1. (E) A baglike section from a fruit containing several seeds; x 1. (F) Seed; x 3.

(G) Seed; x 3. (H) Leaf; x 1/2.

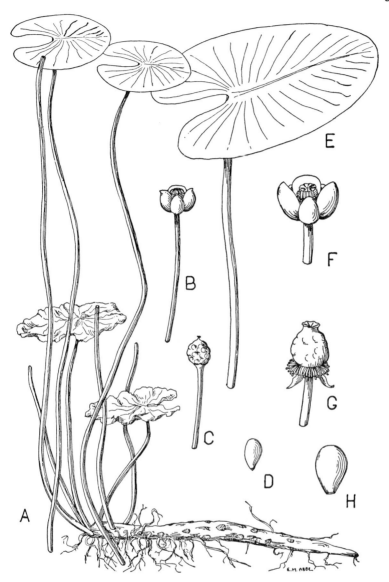

Fig. 106. *Nuphar microphyllum* (A-D) and *N. rubrodiscum* (E-H).

(A) Portion of plant showing submersed leaves and floating leaves; x 1/2. (B) Flower; x 1/2. (C) Fruit; x 1/2. (D) Seed; x 3.
(E) Leaf; x 1/2. (F) Flower; x 1/2. (G) Fruit; x 1/2. (H) Seed; x 3.

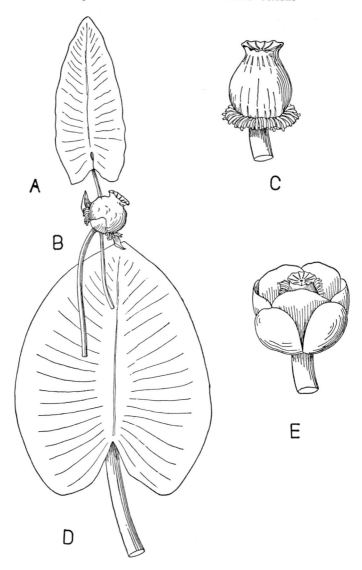

Fig. 107. *Nuphar sagittifolium* (A, B) and *N. polysepalum* (C-E).
(A) Leaf; x 1/4. (B) Fruit; x 1/2.
(C) Fruit; x 1/2. (D) Leaf; x 1/4. (E) Flower; x 1/2.

Map 255. *Nuphar polysepalum.*

Map 256. *Nuphar rubrodiscum.*

Map 257. *Nuphar sagittifolium.*

Map 258. *Nuphar variegatum.*

Nuphar sagittifolium Pursh. Fig. 107, A–B. Map 257.

In sluggish streams, backwaters, and ponds. Along the South Atlantic Coast.

Nuphar variegatum Engelm. (*N. americana* Prov.). Fig. 105, G–H. Map 258.

In shallow ponds and slow streams. Mostly in the northeastern states and the Great Lakes region. Perhaps this is only a variety of *N. advena.*

NYMPHAEA (*CASTALIA* of authors): WHITE WATER LILY

Perennial herbs with erect or creeping rootstocks or tuberous rhizomes; leaves alternate, with slender petiole and floating blade; blade 5 to 60 cm. in diameter, nearly orbicular, entire or slightly crenate, with main veins radiating from the top of the petiole and much branched. Flowers solitary, 2.5 to 30 cm. in diameter, on slender peduncle, usually floating; sepals 4 (rarely 3 or 5), usually green on the outside; petals 12 to 40, white, yellow, pink, or blue, inserted on the ovary; stamens 20 to 700 inserted around the ovary, the outer filaments broad or petaloid, the inner filiform; carpels 8 to 35, fused, and surrounded by and fused laterally with a fleshy, cup-shaped receptacle. Fruit a spongy berry bursting irregularly or decomposing to discharge the several seeds, each with a floating, saclike aril.

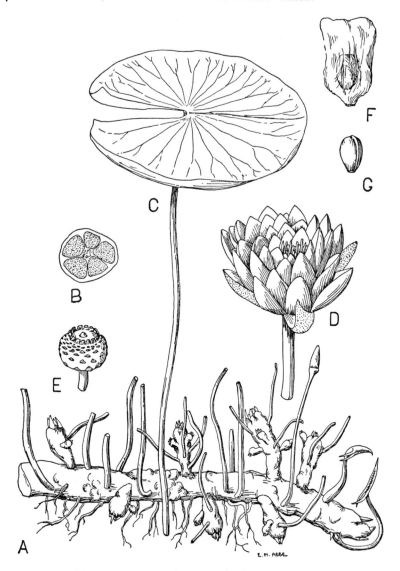

FIG. 108. *Nymphaea tuberosa.*

(A) Rootstock with tubers; x 1/4. (B) Diagram of cross section of rootstock; x 1/2. (C) Peltate leaf; x 1/4. (D) Flower; x 1/2. (E) Fruit; x 1/2. (F) Seed in thin sac; x 3. (G) Seed; x 3.

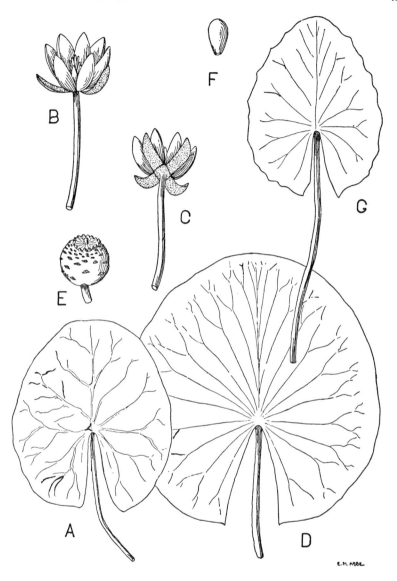

FIG. 109. *Nymphaea tetragona* (A–C), *N. odorata* (D–F), and *N. flava* (G). (A) Leaf; x 1/2. (B) Flower; x 1/2. (C) Flower showing 4 sepals; x 1/2. (D) Leaf; x 1/2. (E) Fruit; x 1/2. (F) Seed; x 3. (G) Leaf; x 1/2.

KEY TO SPECIES OF NYMPHAEA

1. Carpels free at the sides, walls between the cells of the ovary double; petals pale violet .... *N. elegans*
1. Carpels fused at the sides, walls between cells of ovary single.
    2. Receptacle 4-angled, with 4 swollen ridges; rhizome erect.
        3. Flowers yellow, 6 to 13 cm. in diameter; tubers in clusters .... *N. flava*
        3. Flowers white, 3 to 8 cm. in diameter; tubers lacking .... *N. tetragona*
    2. Receptacle not 4-angled; rhizome horizontal; petals white or pink.
        4. Petals spatulate; rhizome with tubers; leaves usually not purple beneath; flowers with little or no scent .... *N. tuberosa*
        4. Petals elliptic; rhizome without tubers; leaves usually purple beneath; flowers sweet-scented .... *N. odorata*

Nymphaea elegans Hook. Blue water lily. Map 259.

In shallow ponds and streams. Southern Texas.

Nymphaea flava Leitn. (*N. mexicana* Zucc.). Yellow lotus, Banana water lily. Fig. 109, G. Map 260.

In shallow ponds and sluggish streams. Mostly along the Gulf Coast from Florida to Texas.

MAP 259. *Nymphaea elegans.*

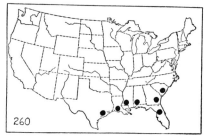

MAP 260. *Nymphaea flava.*

MAP 261. *Nymphaea odorata.*

MAP 262. *Nymphaea tuberosa.*

Nymphaea odorata Ait. Sweet-scented white water lily. Fig. 109, D-F. Map 261.

Common in shallow ponds and lakes. Throughout the eastern states.

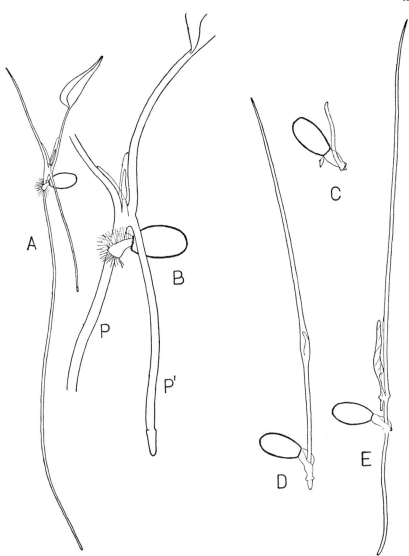

FIG. 110. Seedlings of *Nymphaea tuberosa* (A, B) and *Nuphar advena* (C-E).

(A) Seedling; x 2. (B) Seedling showing primary root (P) and adventitious root (P¹); x 3.

(C) Germinating seed; x 3. (D) Seedling; x 3. (E) Seedling showing primary root, beginning adventitious root, and first true leaf; x 3.

Nymphaea tetragona Georgi. Dwarf water lily. Fig. 109, A–C. Map 263.
Rare in shallow ponds and streams. Maine to Washington.

Nymphaea tuberosa Paine. White water lily. Figs. 108; 110, A–B. Map 262.

Local in bays of the larger lakes and quiet coves in rivers. Chiefly in the upper Mississippi Valley and about the Great Lakes; also introduced locally.

MAP.263. *Nymphaea tetragona.*

REFERENCES

Conard, H. S. The waterlilies: A monograph of the genus *Nymphaea*. Carnegie Inst. Wash. Publ. 4. 1–279. 30 pl., 82 figs. 1905.

Miller, G. S., Jr., and P. C. Standley. The North American species of *Nymphaea* [*Nuphar*]. Contr. U. S. Nat. Herb. 16:63–108. 1912.

Porsild, A. E. *Nymphaea tetragona* Georgi in Canada. Canadian Field Nat. 53: no. 4, 48–50. 1939.

Jones, J. A. Overcoming delayed germination of *Nelumbo lutea*. Bot. Gaz. 85:341–343. 1938.

Ohga, I. The germination of century-old and recently harvested Indian lotus fruits with special reference to the effect of oxygen supply. Amer. Jour. Bot. 13:754–759. 1926.

RANUNCULACEAE: BUTTERCUP FAMILY

Mostly herbs with alternate or, rarely, opposite, palmately lobed, dissected, or compound leaves; petioles mostly expanded at base into stipule-like appendages. Flowers regular or irregular, solitary or in clusters. Sepals 3 to 15, separate; petals 2 to 5, separate; stamens numerous, free; carpels numerous, few or rarely 1; ovary superior, in fruit forming an achene, follicle, or berry. Seeds with endosperm and small embryo. — Mostly terrestrial plants. The following genus is represented by several aquatic species.

RANUNCULUS: Buttercup, Crowfoot

Annuals or perennials with alternate leaves on erect or creeping stems or in basal clusters. Blades palmately dissected or lobed or, rarely, lanceolate to linear and entire. Flowers regular, mostly solitary on a peduncle or a few in a corymb. Sepals 3 to 5; petals mostly 5 to 7 or, rarely, absent, yellow or white; stamens numerous; carpels numerous. Fruits achenes borne in clusters, variously beaked. — About 250 species, only a few aquatics.

KEY TO SPECIES OF RANUNCULUS

1. Flowers yellow; leaves undivided or, if divided, the segments flat.
   2. Leaves, at least the submersed, finely dissected into flat segments; floating leaves, if present, palmately lobed.
      3. Flower 1.5 to 2.5 cm. in diameter; petals much exceeding the sepals; achene with long beak .... *R. flabellaris*
      3. Flower about 1 cm. in diameter; petals not much longer than the sepals; beak short .... *R. purshii*
   2. Leaves entire, denticulate, or crenate, none of them with dissected blades.
      4. Leaves cordate to reniform, with crenate margin, in clusters at the root or nodes .... *R. cymbalaria*
      4. Leaves lanceolate to linear, spatulate, or oblong; entire or denticulate.
         5. Stems ascending, rooted at the lower nodes; petals about as long as the sepals; achene with long beak .... *R. ambigens*
         5. Stems creeping, usually rooted at all the nodes; petals longer than the sepals; achene with short beak .... *R. flammula*
1. Flowers white; leaves mostly on submersed stems, finely dissected, the segments not flat or, if emersed, then the leaves with reniform or palmately lobed blades; flowers solitary on peduncles opposite the leaves.
   6. Leaves mostly submersed and dissected into filiform lobes.
      7. Petiole sheath extending to base of blade; leaves rigid even out of water; receptacle hispid .... *R. circinatus*
      7. Petiole sheath not as long as the petiole; leaves limp and collapsing when removed from water; receptacle hispid or rarely glabrous ..... *R. aquatilis*
   6. Leaves often mostly floating and 3-parted, the segments obcordate or with 3 rounded lobes; receptacle glabrous.
      8. Sepals persistent; submersed leaves finely dissected .... *R. lobbii*
      8. Sepals deciduous; submersed leaves like the floating leaves, merely lobed .... *R. hederaceus*

Ranunculus ambigens L. Water-plantain crowfoot. Fig. 111, A–B. Map 264.

Muddy ponds, streams, and ditches. Northeastern states.

Ranunculus cymbalaria Pursh. Seaside buttercup. Map 268.

In brackish streams and marshes near the seashore; inland also in brackish and alkaline ponds and springs. Widespread.

Ranunculus flabellaris Raf. (*R. delphinifolius* Torr.). Yellow water-crowfoot. Fig. 111, C–D. Map 265.

Mostly submersed in shallow ponds and pools; often left stranded as the water recedes. Northeastern states and Pacific Northwest.

MAP 264. *Ranunculus ambigens*.

MAP 265. *Ranunculus flabellaris*.

Ranunculus flammula L. Spearwort. Fig. 111, E–G. Map 270.

On gravelly and sandy shores of lakes and streams. In clear lakes it may grow submersed to a depth of several meters, the creeping stems rooting at the nodes and forming dense mats over large areas. Such plants are sterile. Widespread in the northeast and from the Rocky Mountains westward.

NOTE. Typical specimens of this species with the leaves mostly from oblanceolate to ovate-lanceolate are common in Europe but infrequent in the United States except in Washington and Oregon. Most of the American material has slenderer leaf blades and has been referred to *R. flammula* var. *reptans* or *R. flammula* var. *filiformis*. These do not seem to be distinct if judged by the variable leaf shapes and large number of intermediate forms.

MAP 266. *Ranunculus hederaceus*.

MAP 267. *Ranunculus lobbii*.

Ranunculus purshii Richards. Fig. 112, G–H. Map 269.

In temporary ponds, sloughs, and ditches, and on muddy shores. From the Rocky Mountains to the Northwest; rare eastward.

Ranunculus aquatilis L. (*Batrachium aquatile* [L.] Wimm.; *R. trichophyllus* Chaix.). White water-crowfoot. Figs. 112, A–D; 102a, E. Map 272.

In ponds, lakes, and streams. Widely distributed except in the southeastern states and the lower Mississippi Valley.

NOTE. This rather polymorphic species has been segregated into several varieties based largely upon the relative hairiness of the receptacle, a character that is very variable, and upon the structure of the leaves, whether they are always all dissected into filiform segments or whether the upper leaves may at times be merely lobed and floating as in var. *hispidulus*. Var. *calvescens* has glabrous or sparsely hairy receptacles; var. *capillaceus* and var. *hispidulus* have a densely hairy receptacle.

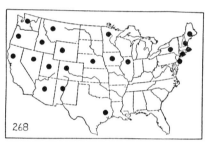

MAP 268. *Ranunculus cymbalaria*.

Ranunculus circinatus Sibth. (*R. longirostris* Godr.; *Batrachium circinatum* Spach.). White water-crowfoot. Fig. 112, E–F. Map 271.

In ponds, shallow lakes, and streams, often forming a dense growth in rather swift brooks. Widespread except in the southern states.

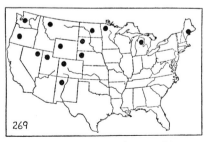 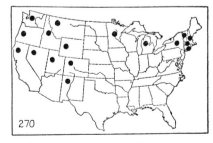

MAP 269. *Ranunculus purshii*.   MAP 270. *Ranunculus flammula*.

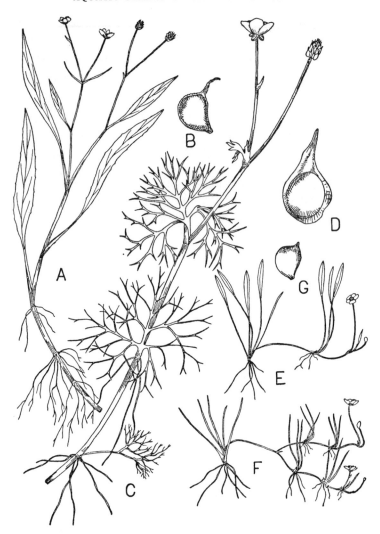

Fig. 111. *Ranunculus ambigens* (A, B), *R. flabellaris* (C, D), and *R. flammula* (E–G).
(A) Habit sketch of plant; x 1/4. (B) Achene; x 5.
(C) Plant; x 1/2. (D) Achene; x 5.
(E) Plant; x 1/2. (F) Plant with narrow leaves; x 1/2. (G) Achene; x 5.

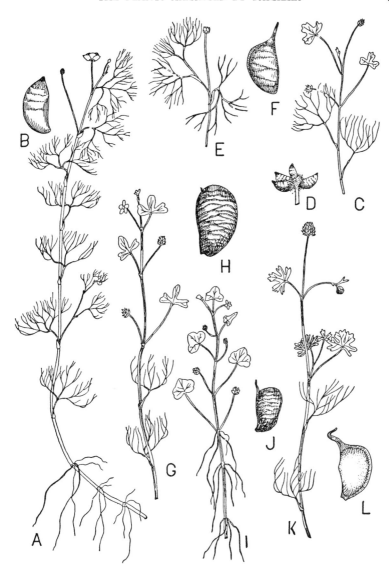

FIG. 112. *Ranunculus aquatilis* (A–D), *R. circinatus* (E, F), *R. purshii* (G, H), *R. lobbii* (I, J), and *R. hederaceus* (K, L).

(A) Plant. (B) Achene. (C) Part of plant with 2 kinds of leaves. (D) Receptacle with achenes; x 2.
(E) Part of plant. (F) Achene.
(G) Part of plant. (H) Achene.
(I) Plant. (J) Achene.
(K) Part of plant. (L) Achene.
Plants all x 1/2. Achenes all x 10.

MAP 271. *Ranunculus circinatus*.

MAP 272. *Ranunculus aquatilis*.

Ranunculus hederaceus L. Fig. 112, K–L. Map 266.

In shallow ponds, streams, and fresh-water marshes. Middle Atlantic Coast.

Ranunculus lobbii Gray. Fig. 112, I–J. Map 267.

In ponds, pools, and sluggish streams. Pacific Coast.

REFERENCES

Benson, Lyman. Pacific States Ranunculi. I–II. Amer. Jour. Bot. 23:26–33, 169–176. 1936.

——. North American Ranunculi. IV–V. Torrey Bot. Club Bull. 69:298–316, 373–386. 1942.

Drew, W. B. The North American representatives of *Ranunculus* (*Batrachium*). Rhodora 38:1–47. 1936.

Lindberg, Harold. *Ranunculus salsuginosus* Pallas (*R. cymbalaria* Pursh.). Bot. Notiser 678–690. 1939.

CRUCIFERAE: Mustard Family

Herbs with alternate leaves and a very pungent or peppery flavor. Flowers in racemes or panicles, perfect, regular. Sepals 4, separate and falling early; petals 4, separate, arranged like a cross; stamens 6, separate, 4 long and 2 short; pistil solitary, with a 2-celled, superior ovary maturing into a capsule or pod which usually dehisces by 2 valves; each valve with 1 or several seeds; seeds with folded embryo, without endosperm. — A large family of terrestrial plants; only a few aquatics.

KEY TO GENERA

1. Leaves all basal, entire, awl-shaped; capsule ovoid or globular .... *Subularia*
1. Leaves cauline, compound, dissected, or serrate.
   2. Leaves pinnately compound; capsule elongated, cylindrical; plant rooting at nodes, often forming floating mats .... *Nasturtium*
   2. Leaves pinnately divided, or with serrate margins if emersed; capsule oval; plant erect .... *Rorippa*

NASTURTIUM: Cress

Perennials with spreading or creeping, angular stems rooting at the nodes. Leaves alternate, pinnately compound with 3 to 11 round or elliptical, nearly entire leaflets. Flowers in racemes, white, on slender pedicels. Capsule linear, usually curved upward, several-seeded.

Nasturtium officinale R. Br. Water cress. Fig. 113, D. Map 273.

Widespread in cold, spring-fed brooks and ponds, rooted on the bottom or along the shore and growing in floating mats over the surface of the water. Introduced from Europe and grown in cultivation as the "water cress" of commerce. Naturalized almost throughout the United States.

Map 273. *Nasturtium officinale.*   Map 274. *Rorippa aquatica.*

RORIPPA: Cress

Perennials with alternate, pinnate or pinnatifid leaves, rooting freely at the lower nodes; stems sparingly branched, mostly submersed. Flowers in racemes, yellow or white. Capsule linear to ovoid, 2-valved, with several seeds usually arranged in 2 rows in each valve. — Several species, mostly of lowlands and marshes, one occurring in water.

Rorippa aquatica. Lake cress. Figs. 113, A–C; 102a, F. Map 274.

In soft, mucky bottom in lakes, and backwaters and coves of rivers. Local mostly east of the Mississippi River.

Note. The lake cress has white flowers and ovoid, 1-celled capsules. In some localities fruits are rarely formed. Reproduction seems to be most common by vegetative propagation by the leaves. In late summer the lower, finely dissected leaves are abscised and fall to the bottom where they send out new roots and a bud.

SUBULARIA: Awlwort

A small, stemless perennial with a tuft of awl-shaped leaves. Flowers minute, white, in few-flowered racemes on erect scapes 2 to 10 cm.

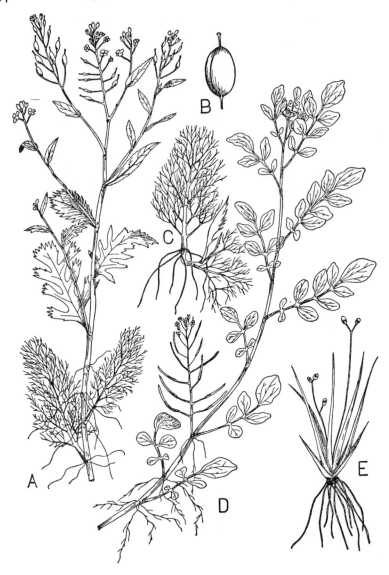

FIG. 113. *Rorippa aquatica* (A–C), *Nasturtium officinale* (D), and *Subularia aquatica* (E).

(A) Portion of flowering plant; x 1/2. (B) Mature capsule; x 2. (C) Abscised leaf showing roots and new shoot; x 1/2.
(D) Portion of plant; x 1/4.
(E) Plant; x 1.

high. Pistil without style, forming an ovoid or globular capsule with a few small seeds. Fruiting rarely except when emersed or exposed. — Only 1 species.

Subularia aquatica L. Fig. 113, E. Map 275.
Submersed in clear, cold lakes or slow streams, usually on sandy

MAP 275. *Subularia aquatica.*

bottom, exposed on shores as the water recedes. Northeastern states and from the northern Rocky Mountains westward.

REFERENCES

Beattie, J. H. Production of water cress. U. S. Dept. Agr. Leaflet 134. 1938.
Muenscher, W. C. Lake-cress. [Propagation by leaves.] In, Aquatic Vegetation of the Champlain Watershed. N. Y. State Dept. Conserv. Suppl. 19th Ann. Rept. 164–185. 1929(1930).

PODOSTEMACEAE: Riverweed Family

Much-branched aquatics anchored on bedrock and stones in swift water. Cartilaginous, appearing like marine algae. Flowers very small, naked, surrounded by an involucre. Fruit a 2- to 3-celled, many-seeded, capsule. — Widely distributed in tropical streams. Only the following genus represented in North America.

PODOSTEMUM: Riverweed

Plants attached to rocks by fleshy disks or rootlike processes. Leaves 2-ranked, olive-green or reddish, divided into linear lobes or segments. Flowers naked, in a tubular sheath. Pistil with a stalk to 1 side of which are attached 2 fertile stamens with long filaments fused more than half their length. A short, sterile filament attached to each side of the fertile stamens. Ovary stalked, with 2 subulate stigmas. Capsule stalked, 8-ribbed, 2-celled, 2-valved, with many seeds on a thick,

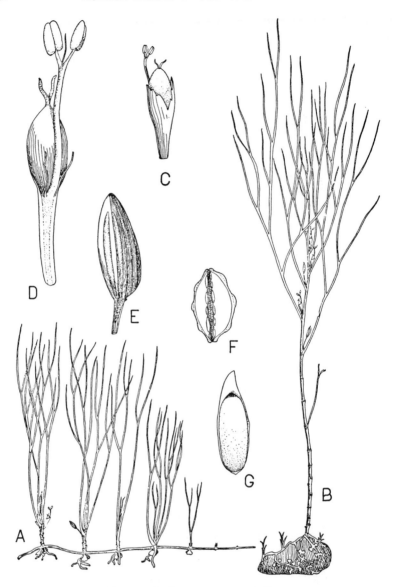

Fig. 114. *Podostemum ceratophyllum.*

(A) Plant showing creeping habit and sucker-like holdfasts; x 1. (B) Flowering plant; x 1. (C) Flower with spathe; x 5. (D) Flower with spathe removed; x 8. (E) Mature capsule; x 5. (F) Cross section of capsule showing placenta and two locules; x 5. (G) Seed; x 50.

central placenta. Seeds minute without endosperm. — About 10 species. Widely distributed.

KEY TO SPECIES OF PODOSTEMUM

1. Flowers solitary; leaf segments flattened .... *P. ceratophyllum*
1. Flowers 2 or 3 together; leaf segments filiform .... *P. abrotanoides*

Podostemum abrotanoides Nutt. Map 277.

On stones and gravel in streams. From Tennessee to northern Florida.

Podostemum ceratophyllum Michx. Fig. 114. Map 276.

On bedrock and stones in swift rivers and brooks. From Maine to Texas. The length of the leaf segments and amount of branching vary with the depth and swiftness of the water.

Map 276. *Podostemum ceratophyllum*.   Map 277. *Podostemum abrotanoides*.

REFERENCES

Muenscher, W. C., and Bassett Maguire. Notes on some New York plants. Rhodora 33:165–167. 1 pl. 1931.

Fassett, N. C. *Podostemum* in North America. Rhodora 41:526–529. 1939.

CRASSULACEAE: Stonecrop Family

Herbs with alternate or opposite, simple, mostly succulent leaves. Flowers perfect, regular, in cymose clusters or axillary. Sepals 3 to 20; petals and carpels of the same number as sepals, the stamens twice as many; parts of each whorl separate or fused only at their base. Fruit a follicle. — A large family of terrestrial plants many of them of deserts; the following genus mostly terrestrial, 1 aquatic species.

TILLAEA

Small, tufted or creeping annuals rooting at the base. Leaves opposite, entire. Flowers axillary, nearly sessile, with 3 or 4 sepals, petals, stamens, and carpels. Follicles with about 10 seeds.

Tillaea aquatica L. Fig. 115, A–C. Map 278.

Local on brackish shores or tidal mud flats in rivers and estuaries. Mostly along the Atlantic Coast.

ROSACEAE: Rose Family

Plants woody or herbaceous, with alternate, stipulate leaves. Flowers regular; sepals and petals usually 5; stamens numerous; carpels 1 to many; ovary and fruit various. — Primarily terrestrial plants.

POTENTILLA: Cinquefoil, Five-Finger

Herbs or shrubs with alternate, compound leaves. Flowers solitary or cymose. Calyx 5-cleft, subtended by 5 bractlets, giving it the appearance of being 10-cleft; petals 5, rounded; stamens numerous; carpels numerous, forming achenes on a hairy receptacle. — A large genus all the species of which are terrestrial except the following which is rarely aquatic.

Potentilla palustris (L.) Scop. Marsh cinquefoil. Fig. 115, D–E. Map 279.

Leaves pinnately compound; petals reddish-purple. Rooted along borders of ponds and streams, mostly in acid bogs. Frequently the rootstocks spread out over the water forming dense mats. Widely distributed across the northern United States except on the Great Plains.

Map. 278. *Tillaea aquatica.*

Map 279. *Potentilla palustris.*

CALLITRICHACEAE: Water-Starwort Family

Small, monoecious, aquatic perennials, or rarely terrestrial annuals, with slender, tufted stems and opposite, entire, linear or spatulate to

Fig. 115. *Tillaea aquatica* (A–C) and *Potentilla palustris* (D, E).
(A) Plants; x 1. (B) Mature carpels; x 10. (C) Seed; x 20.
(D) Plant; x 1/2. (E) Cluster of ripe achenes; x 1/2.

obovate leaves, the latter usually in floating rosettes. Flowers solitary or in groups of 2 or 3 in the leaf axils, small, naked or subtended by a pair of bracts. Staminate flower a single stamen. Pistillate flower a 4-celled ovary with 2 stigmas. Fruit nutlike, flattened, 4-celled, 4-lobed, when ripe separating into four 1-seeded sections. Seeds pendulous, with endosperm.

CALLITRICHE: WATER STARWORT

The only genus; with the characteristics of the family. — About 20 species.

KEY TO SPECIES OF CALLITRICHE

1. Leaves all linear, 1-veined, submersed; mature carpels separating more than ½ their length .... *C. hermaphroditica*
1. Leaves, at least the floating ones, spatulate or obovate, 3-veined; mature carpels not separating.
   2. Fruit higher than wide .... *C. palustris*
   2. Fruit about as high as wide.
      3. Mature fruits 1.5 to 2 mm. high, green; carpels winged on the outer margin; floating leaves dark-green, coarse .... *C. stagnalis*
      3. Mature fruits about 1 to 1.5 mm. high, brown; carpels not winged; floating leaves lighter green, smaller .... *C. heterophylla*

Callitriche hermaphroditica L. (*C. autumnalis* L.). Fig. 116, A–B. Map 281.

In lakes, ponds, and slow streams. Local, mostly in the northern states.

Callitriche heterophylla Pursh. Fig. 116, C–D. Map 280.

In shallow ponds and slow streams. Throughout the eastern states, local on the Pacific Coast. Plants from the Pacific Coast with styles about twice as long as the fruit have been described as *C. bolanderi* Hegelm.

Callitriche palustris L. Fig. 116, E–F. Map 282.

In ponds, pools, and sluggish streams. Throughout the northern states.

Callitriche stagnalis Scop. Fig. 116, G–H. Map 283.

In cold, spring-fed ponds and pools and in swift brooks mostly from springs. Common south of the moraine on Long Island, New York. Local along the North Atlantic Coast; rare inland.

THE PLANTS ARRANGED BY FAMILIES 261

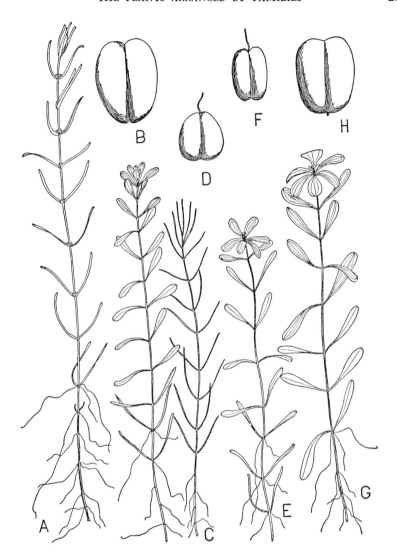

Fig. 116. *Callitriche hermaphroditica* (A, B), *C. heterophylla* (C, D), *C. palustris* (E, F), and *C. stagnalis* (G, H).

(A, C, E, G) Plants; x 1.
(B, D, F, H) Fruits; x 15.

MAP 280. *Callitriche heterophylla*.   MAP 281. *Callitriche hermaphroditica*.

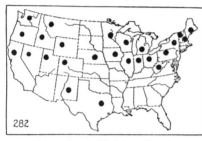

MAP 282. *Callitriche palustris*.   MAP 283. *Callitriche stagnalis*.

REFERENCE

Svenson, H. K. *Callitriche stagnalis* in eastern United States. Rhodora 34:37–38. 1932.

ELATINACEAE: WATERWORT FAMILY

Dwarf annuals with erect or creeping stems. Leaves entire or serrulate, opposite, glandular-dotted. Flowers small, perfect, regular, axillary. Sepals 2 to 5, imbricated; petals 2 to 5; stamens 2 to 10; carpels 2 to 5, fused into a compound pistil with superior ovary. Fruit a capsule with several seeds. — Two genera with about 25 species.

ELATINE: WATERWORT

Aquatics with stems rooting at the nodes and usually partly buried in sand or mud. Leaves entire, mostly obovate or spatulate. Flowers solitary in the leaf axils, sessile or nearly so. Sepals 2 to 4, separate; petals 2 to 4; stamens 2 to 8, separate; pistil with 2- to 4-celled ovary with as many styles or sessile stigmas. Capsule 2- to 4-valved, several- to many-seeded, the seeds vertical and attached to a basal placenta or nearly horizontal and attached all along an axial placenta. Seeds with longitudinal lines and crossbars. — About 8 species, mostly aquatics.

THE PLANTS ARRANGED BY FAMILIES 263

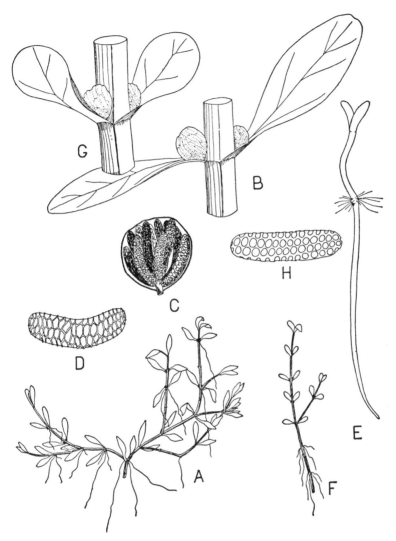

Fig. 117. *Elatine americana* (A–E) and *E. minima* (F–H).

(A) Plant; x 1. (B) Node with axillary fruits; x 10. (C) Capsule in vertical section; x 20. (D) Seed; x 50. (E) Seedling; x 8.
(F) Plant; x 1. (G) Node with fruits; x 10. (H) Seed; x 50.

KEY TO SPECIES OF ELATINE

1. Flowers 2-merous; seeds mostly straight or nearly so, 0.2 to 0.3 mm. thick, with distinct longitudinal ribs and rounded pits; stems simple or sparingly branched .... *E. minima*
1. Flowers 3-merous; seeds with indistinct longitudinal ridges and angular pits; stems freely branching, often prostrate or creeping.
    2. Seeds curved into a semicircle .... *E. californica*
    2. Seeds slightly curved or nearly straight.
        3. Leaves rounded at apex, thick; seeds erect, from a basal placenta .... *E. americana*
        3. Leaves emarginate at apex, thin; seeds nearly horizontal, attached along an axile placenta .... *E. triandra*

Elatine americana (Pursh) Arn. Fig. 117, A–E. Map 284.

Local on wet clay or silty tidal flats and borders of ponds and streams. Widespread across the northern and central states.

Elatine californica Gray. Map 286.

On muddy bottom of ponds and along slow streams. Eastern Washington to Montana and California.

Map 284. *Elatine americana.*

Map 285. *Elatine minima.*

Map 286. *Elatine californica.*

Map 287. *Elatine triandra.*

Elatine minima (Nutt.) Fisch. and Meyer. Fig. 117, F–H. Map 285.

Locally abundant in shallow lakes and ponds and along slow streams, mostly on sandy bottom. Northeastern states and the Great Lakes region.

Elatine triandra Schkuhr. Map 287.

Rare, shores of lakes and ponds. From Maine westward to the Rocky Mountain region; perhaps introduced.

REFERENCES

Fernald, M. L. The genus *Elatine* in eastern North America. Rhodora 19:10–13. 1917.
——. *Elatine americana* and *E. triandra*. Rhodora 43:208–211. 1941.
Fassett, N. C. *Elatine* and other aquatics. Rhodora 41:367–377. 1939.

LYTHRACEAE: Loosestrife Family

Herbs mostly with 4-sided stems and opposite or whorled, entire leaves without stipules. Flowers perfect, mostly regular, all alike or of 2 or 3 kinds, axillary, or in whorls in a spike. Calyx campanulate or tubular, inclosing the other organs. Petals 4 to 7 or none, stamens 4 to 14, all attached to the calyx throat. Pistil solitary, with superior ovary and 1 style with capitate or 2-lobed stigma; fruit a capsule with axile placenta, 1- or 2-celled, many-seeded; seeds without endosperm.

KEY TO GENERA

1. Plants low, slender, spreading, mostly submersed; petals none; stamens 4 .. *Peplis*
1. Plants tall, emersed or in marshes; stem often corky.
    2. Stems erect; calyx tubular; petals 6; stamens 6 or 12 .... *Lythrum*
    2. Stems spreading, stoloniferous; calyx cup-shaped; petals 5; stamens 8 to 10
        .... *Decodon*

DECODON: Swamp Loosestrife

Coarse perennials with angular, recurved stems mostly 1 to 2 meters long, often rooting at the tips and producing spongy bark tissue on the submersed parts. Leaves opposite or whorled, nearly sessile, lanceolate. Flowers on short pedicels, clustered in the axils of the upper leaves, trimorphous. Calyx cup-shaped, with 5 to 7 teeth, with as many slender bracteoles in the sinuses; corolla of 5 (rarely 4) pink or rose petals; stamens 8 to 10, exserted, 4 or 5 long, the others short. Capsule globose, 3- to 5-celled, loculicidal.

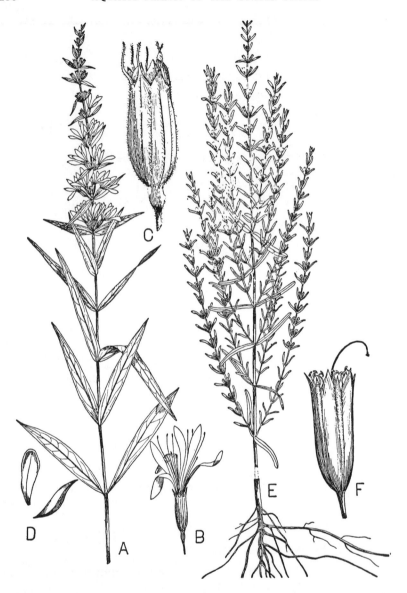

Fig. 118. *Lythrum salicaria* (A-D) and *L. lineare* (E, F).
(A) Flowering shoot; x 1/2. (B) Flower; x 2. (C) Capsule; x 8. (D) Seeds; x 15. (E) Plant; x 1/2. (F) Capsule; x 8.

THE PLANTS ARRANGED BY FAMILIES 267

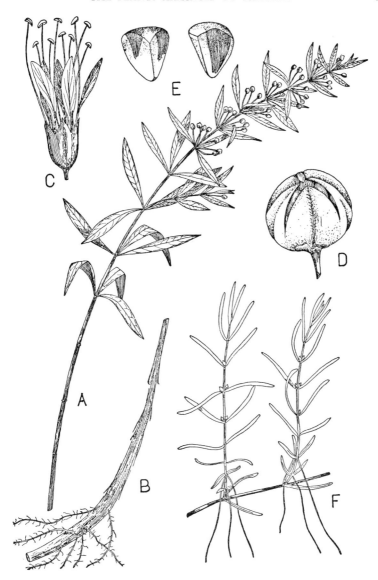

FIG. 119. *Decodon verticillatus* (A–E) and *Peplis diandra* (F).

(A) Flowering branch; x 1/4. (B) Base of plant showing corky stem; x 1/4. (C) Flower; x 2. (D) Capsule; x 4. (E) Seeds; x 10.
(F) Plant; x 1.

Decodon verticillatus (L.) Ell. Swamp Loosestrife, Water willow. Fig. 119, A-E. Map 288.

Common on swampy margins of ponds and streams, usually extending into the water. Throughout the eastern states.

LYTHRUM: LOOSESTRIFE

Slender, erect, hard-stemmed herbs. Leaves opposite or whorled, linear to lanceolate or nearly ovate. Flowers solitary in the upper leaf axils or in whorls on a terminal spike or cluster of spikes. Calyx cylindrical, striate, 5- to 7-toothed, with bractlets in the sinuses; petals 6 (5 to 7), pink, purple, or white; stamens as many as the petals or twice as many; pistil in some flowers with short style, in others with long style. Capsule narrow, 2-celled, many-seeded.

KEY TO SPECIES OF LYTHRUM

1. Leaves linear; flowers small, solitary in upper leaf axils, white; stamens 5 to 7 .... *L. lineare*
1. Leaves lanceolate; flowers large, in spikes, purple; stamens 8 to 12 .. *L. salicaria*

Lythrum lineare L. Fig. 118, E-F. Map 290.

Brackish marshes and pools. Coastal Plain from New Jersey to Florida and Texas.

Lythrum salicaria L. Spiked loosestrife, Purple loosestrife. Fig. 118, A-D. Map 289.

Common along flooded shores of streams and ponds, in sloughs and wet meadows. Naturalized from Eurasia, spreading in recent years.

MAP 288. *Decodon verticillatus.*

MAP 289. *Lythrum salicaria.*

PEPLIS: WATER PURSLANE

Delicate dwarf plants, usually submersed. Leaves linear, sessile from a broad base. Flowers small, green, axillary. Calyx with triangular

MAP 290. *Lythrum lineare.*

MAP 291. *Peplis diandra.*

lobes; corolla wanting; stamens 4; capsule small, globose, 2-celled, indehiscent.

Peplis diandra Nutt. (*Didiplis diandra* [Nutt.] Wood.). Fig. 119, F. Map 291.

In shallow ponds and along borders of lakes and streams. Chiefly in the Mississippi Basin; infrequent in the southeastern states.

ONAGRACEAE: Evening-Primrose Family

Herbs with opposite or alternate, mostly simple leaves. Flowers perfect, regular, solitary in leaf axils or in clusters. Calyx fused with the ovary and extended as a tube above it or terminating in 2 to 6 lobes on its summit, the lobes persistent or deciduous. Petals 2 to 6 or wanting; stamens as many as, or twice as many as, the petals and sepals and attached on top of the calyx tube. Pistil with an inferior, 2- to 4-celled ovary with a slender style and capitate or 2- to 4-lobed stigma. Fruit a many-seeded capsule or 1-seeded and indehiscent. Seeds without endosperm. — Mostly terrestrial plants; a few aquatics.

KEY TO GENERA

1. Petals 4 or wanting; stamens 4; capsule short, nearly globose or short cylindrical .... *Ludvigia*
1. Petals 4 to 6; stamens 8 to 12; capsule elongated, at least twice as long as broad .... *Jussiaea*

JUSSIAEA: Primrose Willow

Perennial herbs with alternate or opposite leaves. Calyx lobes 4 to 6, persistent; petals 4 to 6, mostly yellow and showy; stamens 8 to 12. Capsule elongated, 4- to 6-celled. — About 50 species, mostly of tropical regions.

KEY TO SPECIES OF JUSSIAEA

1. Stems erect; petals 4.
   2. Leaves sessile, glabrous .... *J. decurrens*
   2. Leaves petioled, hairy .... *J. peruviana*
1. Stems creeping, rooted or floating; petals 5.
   3. Leaves and stem hairy; corolla 4 to 5 cm. wide .... *J. grandiflora*
   3. Leaves and stem glabrous; corolla 2 to 3 cm. wide .... *J. diffusa*

Jussiaea decurrens (Walt.) DC. Map 292.

In marshes, sloughs, and shallow water of ponds and streams. Southeastern states.

Jussiaea diffusa Forsk. Fig. 120, C, E–F. Map 294.

In ponds and on banks of streams. Mostly in the lower Mississippi Basin. The California plants, *J. californica* (Wats.) Jepson, seem to be indistinguishable from this species.

Jussiaea grandiflora Michx. Fig. 120, D. Map 293.

In marshes and borders of ponds and streams. Coastal Plain from North Carolina to Florida and Louisiana.

Jussiaea peruviana L. Fig. 120, A–B. Map 295.

In shallow ponds and along shores of lakes and streams. Florida and Louisiana.

Map 292. *Jussiaea decurrens.*

Map 293. *Jussiaea grandiflora.*

Map 294. *Jussiaea diffusa.*

Map 295. *Jussiaea peruviana.*

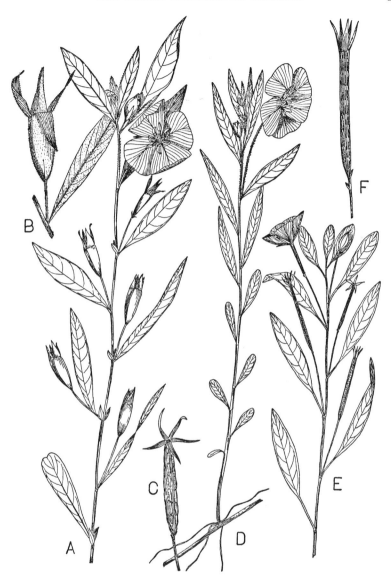

Fig. 120. *Jussiaea peruviana* (A, B), *J. diffusa* (C, E, F), and *J. grandiflora* (D).
(A) Upper part of plant; x 1/2. (B) Capsule; x 1.
(C) Capsule; x 1. (E) Flowering branch; x 1/2. (F) Capsule; x 1.
(D) Plant; x 1/2.

LUDVIGIA: False Loosestrife

Perennial herbs with opposite or alternate leaves and axillary flowers. Calyx with 4 persistent lobes; petals 4 or none; stamens 4. Capsule short or cylindrical, many-seeded, loculicidal.

KEY TO SPECIES OF LUDVIGIA

1. Leaves opposite, petioled; stems creeping and rooting at the nodes, or floating; petals wanting; capsules sessile, axillary .... *L. palustris*
1. Leaves alternate, mostly sessile or nearly so; stems mostly erect (except for sterile runners).
   2. Flowers on pedicels; petals showy, yellow; capsule nearly cubical .... *L. alternifolia*
   2. Flowers sessile; petals small, greenish, or wanting.
      3. Capsule about as long as broad; stems tall, with spongy bark at base .... *L. sphaerocarpa*
      3. Capsule much longer than broad; stem low .... *L. linearis*

Ludvigia alternifolia L. Seedbox. Fig. 121, C–D. Map 296.

In marshes and along borders of streams and ponds. Widespread throughout the eastern states and lower Mississippi Valley.

MAP 296. *Ludvigia alternifolia.*

MAP 297. *Ludvigia linearis.*

MAP 298. *Ludvigia palustris.*

MAP 299. *Ludvigia sphaerocarpa.*

THE PLANTS ARRANGED BY FAMILIES 273

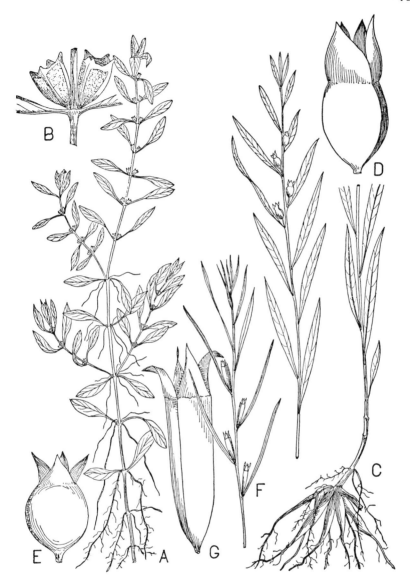

FIG. 121. *Ludvigia palustris* (A, B), *L. alternifolia* (C, D), *L. sphaerocarpa*, (E), and *L. linearis* (F, G).

(A) Plant; x 1/2. (B) Fruits; x 5.
(C) Plant; x 1/2. (D) Capsule; x 5.
(E) Capsule; x 5.
(F) Flowering branch; x 1/2. (G) Capsule; x 5.

Ludvigia linearis Walt. Fig. 121, F–G. Map 297.

In marshes and along borders of streams and ponds. From the Gulf Coast northward along the Atlantic Coast.

Ludvigia palustris (L.) Ell. Water purslane. Fig. 121, A–B. Map 298.

Common in shallow ponds, streams, and marshes. Throughout the eastern states and local westward.

NOTE. A variable species, often growing completely submersed in swift spring brooks. In shallow ponds it may form floating mats. On wet shores it may root in the mud and form a carpet-like mat of creeping stems.

Ludvigia sphaerocarpa Ell. Fig. 121, E. Map 299.

In shallow ponds, lakes, and marshes. Sometimes it grows in floating mats that rise and fall with the water level. Chiefly along the Atlantic and Gulf Coasts.

TRAPACEAE: WATER-NUT FAMILY

Aquatic herbs with long, cordlike, sparsely branching, submersed stems. Leaves of 3 kinds: linear, mostly alternate along the length of the stem, early deciduous; a pair of branched, plumelike structures up to 8 cm. in length at each node; a rosette of floating leaves with inflated petioles and expanded blades terminating each stem or branch. Flowers perfect, on short, thick, axillary peduncles, projecting above the water surface. Calyx tube short, the limb 4-parted, the lobes persistent, all 4 or only the 2 laterals forming spines. Petals 4, sessile, inserted on the perigynous disk; stamens 4, inserted with the petals. Pistil with a 2-celled ovary, subulate style, and capitate stigma. Fruit a nut with 2 to 4 spines, or swollen at the middle, 1-celled, 1-seeded. — A single genus of about 5 species native to Europe, tropical Africa and Asia.

TRAPA: WATER NUT, WATER CHESTNUT, CALTROPS

Annuals with stems from 2 to 5 meters long. Rosette leaves with rhombic blades coarsely serrate, the upper surface glossy and the lower hairy. Nut with 2 or 4 sharp prongs, surrounded by receptacle and calyx except near the summit, which is open and encircled by a crown of bristles. The mature nuts sink to the bottom and after shedding the outer layer have a black or metallic appearance; the stout prongs then reveal sharp, retrorse spines. Seed angular, cordate, without endosperm; embryo with very unequal cotyledons, one a small scale, the other large and fleshy with starch reserves (Fig. 123). The seeds remain viable for 1 winter only. If they are air-dried they die.

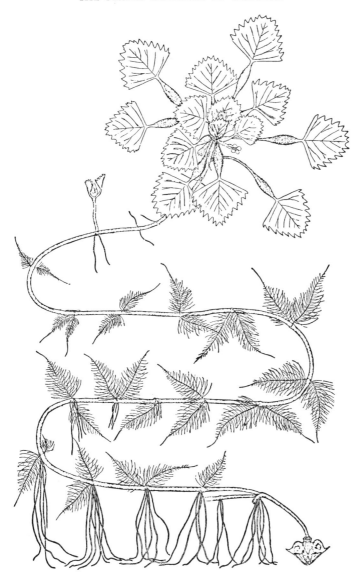

Fig. 122. *Trapa natans*.

A plant 9 weeks after seed germination. Note the pair of plumelike structures at each node of the stem. The unbranched adventitious roots from the lower nodes grow downward into the muddy substrate. (After Winne, 1935.)

Trapa natans L. Water nut, Water chestnut. Fig. 122. Map 300.

Rooting in soft mud in lakes, ponds, canals, and slow backwaters and bays of rivers, in from 1 to 5 meters of water. Native to the Old World. Introduced locally in the eastern United States; established as a weed in the Mohawk and Hudson Rivers in New York, the Sudbury River in Massachusetts, and in the lower part of the Potomac River between Maryland and Virginia.

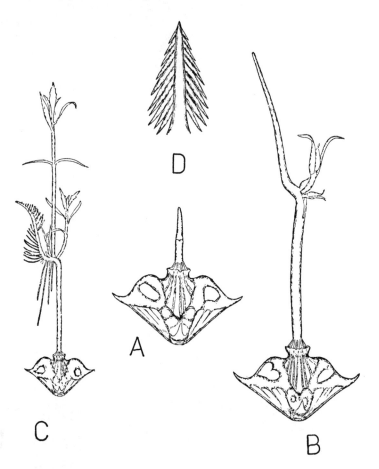

FIG. 123. Seedling stages of *Trapa natans*. (A) Nut with protruding seedling 4 days after germination, showing hypocotyl and first cotyledon stalk; x 1. (B) Showing elongation (end of second week) of first cotyledon stalk and hypocotyl (pointing upward); x 1; note the flaplike second cotyledon beneath which the first and second shoots protrude. (C) Seedling (end of third week); x 1/2; secondary roots have formed from the hypocotyl, those nearest the primary node being positively geotropic. (D) Apex of spine; highly magnified. (After Winne, 1935.)

REFERENCES

Muenscher, W. C. The water chestnut. In, A Biological Survey of the Mohawk-Hudson Watershed. New York State Dept. Conserv. Suppl. Ann. Rept. IX. 234–242. 1934 (1935).

Winne, W. T. A study of the water chestnut, *Trapa natans*, with a view to its control in the Mohawk River. Cornell Univ. Thesis. 1935.

Map 300. *Trapa natans.* Map 301. *Hippuris vulgaris.*

HALORAGIDACEAE: WATER-MILFOIL FAMILY

Perennials with lax and submersed, creeping, or emersed stems. Leaves alternate or whorled, and finely dissected, entire, or serrate, frequently variable on the same plant. Flowers sessile in the axils of the leaves or bracts, regular, perfect or imperfect, often much reduced. Calyx fused with the ovary, its limb very short or wanting; corolla reduced or wanting; stamens 1 to 8. Carpels solitary, or 2 to 4 more or less united into a compound pistil with distinct styles or sessile stigmas. Fruit indehiscent, with 1 to 4 cells each 1-seeded. Seeds with fleshy endosperm.

KEY TO GENERA

1. Leaves entire, whorled; flowers with a solitary stamen and pistil ...... *Hippuris*
1. Leaves not entire, the submersed ones usually dissected, divided, or serrate.
   2. Flower parts in 3's; leaves alternate .... *Proserpinaca*
   2. Flower parts in 4's; leaves mostly whorled (in some species alternate) .... *Myriophyllum*

HIPPURIS: MARE'S-TAIL

Perennial aquatics with forked, jointed, creeping rootstocks with roots and scales at the nodes. Stems simple, submersed and lax, or emersed and erect, with whorls of 6 to 12 leaves at the nodes. Leaves slender and flaccid on submersed stems or firm and linear-acute on emersed stems. Flowers perfect or imperfect. Calyx without teeth;

petals wanting; stamen and carpel solitary; fruit an oblong nutlet, about 2 mm. long, 1-seeded. — A few species of wide distribution but only the following is represented in the waters of the United States.

Hippuris vulgaris L. Fig. 124. Map 301.

In rivers, streams, ditches, and shallow, marshy ponds. In the St. Lawrence River the submersed form is common in swiftly flowing channels between islands. Widespread in the northern states.

MYRIOPHYLLUM: Water Milfoil

Perennial aquatics with slender, sparingly branched stems mostly rooting freely at the lower nodes. Leaves whorled or alternate, variable, the lower usually pinnately dissected into filiform segments, the upper often reduced to bracts. Flowers perfect or imperfect, sessile in the axils of the upper emersed leaves (floral bracts) or in the axils of submersed foliage leaves; the uppermost flowers staminate. Calyx 4-parted in the staminate flowers or 4-toothed in the pistillate flowers. Corolla of 4 petals or wanting. Stamens mostly 4 or 8. Carpels 4, fused into a 4-celled pistil. Fruit deeply 4-lobed; each lobe 1-seeded like a nutlet, rounded or with a ridge or row of tubercles on the back. — About 20 species.

KEY TO SPECIES OF MYRIOPHYLLUM

1. Leaves simple, reduced to small, blunt scales; stems erect and crowded, from creeping rootstocks .... *M. tenellum*
1. Leaves dissected into narrow segments.
    2. Leaves (except sometimes the floral bracts) all whorled.
        3. Flowers in the axils of submersed leaves .... *M. brasiliense*
        3. Flowers in an open or close spike.
            4. Flower-bearing bracts alternate .... *M. alterniflorum*
            4. Flower-bearing bracts in whorls.
                5. Bracts pinnately dissected or lobed; stamens 4 or 8 .... *M. verticillatum*
                5. Bracts entire or toothed.
                    6. Bracts oval, not longer than the fruit; stamens 8 ..... *M. exalbescens*
                    6. Bracts oblanceolate, toothed, much longer than the fruit; stamens 4
                    .... *M. heterophyllum*
    2. Leaves, at least some of them, alternate (sometimes partly whorled).
        7. Fruiting carpels without ridges or tubercles on the back, rounded; flowers borne in the axils of emersed leaves with flattened segments or in the axils of submersed leaves with filiform segments .... *M. humile*
        7. Fruiting carpels with a ridge or row of tubercles on the back.
            8. Leaves, including those with flowers in their axils, all alike, submersed
            .... *M. farwellii*
            8. Leaves of 2 kinds, the submersed with filiform segments, the floral leaves with flat segments on emersed stems .... *M. scabratum*

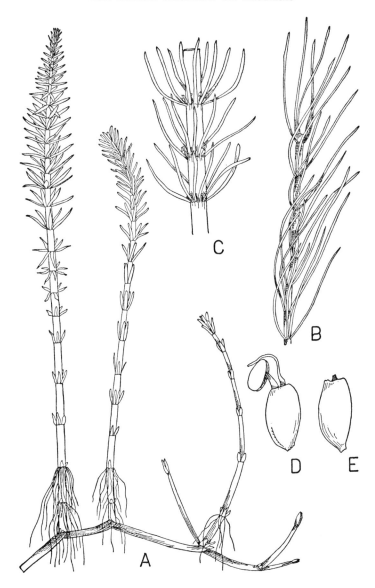

Fig. 124. *Hippuris vulgaris.*

(A) Habit sketch showing young shoots from creeping rootstocks; x 1/3. (B) Portion of submersed shoot with long leaves; x 1/2. (C) Portion of shoot showing flowers in leaf axils; x 1. (D) Flower; x 8. (E) Fruit; x 8.

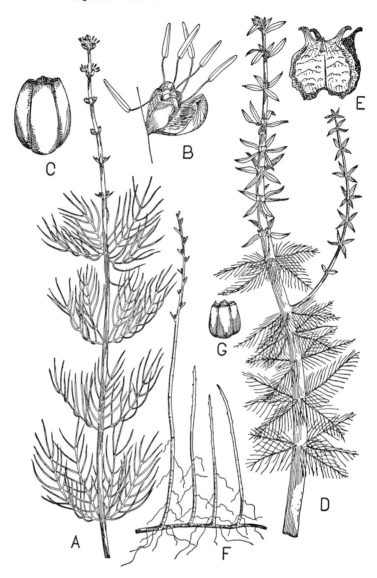

Fig. 125. *Myriophyllum exalbescens* (A–C), *M. heterophyllum* (D, E), and *M. tenellum* (F, G).

(A) Upper part of plant; x 1. (B) Staminate flower; x 10. (C) Group of 4 nutlets; x 10.
(D) Upper part of plant; x 1. (E) Nutlets; x 10.
(F) Plant; x 1. (G) Nutlets; x 10.

THE PLANTS ARRANGED BY FAMILIES 281

Fig. 126. *Myriophyllum verticillatum* (A, B), *M. scabratum* (C–E), and *M. brasiliense* (F).

(A) Part of plant; x 1. (B) Nutlets; x 10.
(C) Part of aquatic plant; x 1. (D) Terrestrial form; x 1. (E) Nutlets; x 10.
(F) Upper part of plant; x 1.

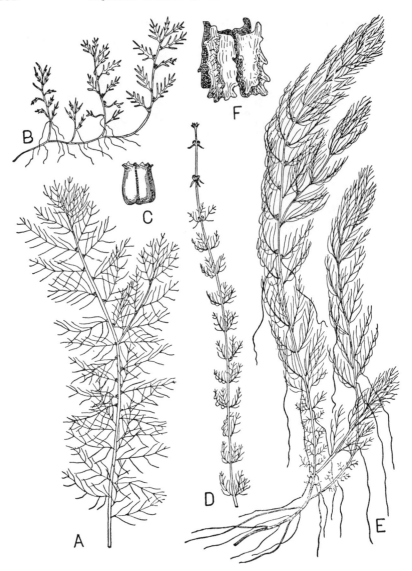

FIG. 127. *Myriophyllum humile* (A–C), *M. alterniflorum* (D), and *M. farwellii* (E, F). (A) Part of aquatic plant; x 1. (B) Part of terrestrial form of plant; x 1. (C) Nutlets; x 10.
(D) Upper part of plant; x 1.
(E) Plant; x 1. (F) Nutlets; x 10.

Myriophyllum alterniflorum DC. Fig. 127, D. Map 302.

Common in shallow water of rivers and lakes; mostly over limestone. Northeastern and Great Lakes states.

Myriophyllum brasiliense Comb. (*M. proserpinacoides* Gill.). Fig. 126, F. Map 307.

Locally common in spring-fed brooks on Long Island, New York; infrequent elsewhere. Naturalized from Brazil.

NOTE. This is the "parrot's feather" of the dealer in aquarium plants. It is a common subject in aquaria and summer pools where it may grow emersed.

Myriophyllum exalbescens Fernald. Figs. 125, A–C; 102a, B. Map 303.

Common in 1 to 4 meters of water in lakes and ponds and near the mouths of streams. Widespread, chiefly in calcareous regions and mostly in the northern states.

MAP 302. *Myriophyllum alterniflorum.*

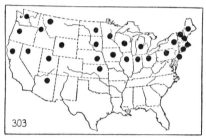
MAP 303. *Myriophyllum exalbescens.*

MAP 304. *Myriophyllum farwellii.*

MAP 305. *Myriophyllum heterophyllum.*

Myriophyllum farwellii Morong. Fig. 127, E–F. Map 304.

Locally common in shallow, acid ponds with organic or sandy bottom. New England states and northern New York, infrequent in the Great Lakes region.

Myriophyllum heterophyllum Michx. Fig. 125, D–E. Map 305.

In ponds, lakes, and sluggish streams. Nearly throughout the United States.

MAP 306. *Myriophyllum humile*.

MAP 307. *Myriophyllum brasiliense*.

MAP 308. *Myriophyllum scabratum*.

MAP 309. *Myriophyllum tenellum*.

Myriophyllum humile (Raf.) Morong. Fig. 127, A–C. Map 306.

A variable species with filiform leaf segments when growing submersed or with broad leaf segments when emersed. In shallow ponds and streams or on muddy banks after the water recedes. Chiefly in the northeastern states.

Myriophyllum scabratum Michx. Fig. 126, C–E. Map 308.

In streams and ponds. Chiefly in the Mississippi Valley.

Myriophyllum tenellum Bigel. Fig. 125, F–G. Map 309.

Common in acid ponds, lakes, and streams, mostly on sandy soil. New England and New York; local about the Great Lakes.

Myriophyllum verticillatum L. Fig. 126, A–B. Map 310.

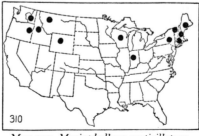

MAP 310. *Myriophyllum verticillatum*.

In deep water of lakes and rivers; best developed along the edges of currents made by rivers or streams entering lakes. Widespread and local in the northern states.

PROSERPINACA: Mermaid Weed

Low perennials with simple or sparsely branched stems with creeping and rooting base. Leaves alternate, pinnately dissected, or the upper lanceolate and serrate, those on the same plant uniform or of both extreme types and intermediate forms. Flowers sessile and solitary in the leaf axils, perfect. Calyx 3-parted; corolla wanting; stamens 3; pistil 3-angled, with 3 stigmas. Fruit nutlike, 3-angled, 3-celled, 3-seeded. — Two or 3 species; North American.

KEY TO SPECIES OF PROSERPINACA

1. Upper leaves lanceolate, serrate .... P. palustris
1. Upper leaves like the lower, pinnately divided .... P. pectinata

Proserpinaca palustris L. Fig. 128, A–C. Map 312.

In sloughs, shallow ponds, and temporary pools, frequently in shaded situations. Widespread throughout the eastern United States.

Proserpinaca pectinata Lam. Fig. 128, D. Map 311.

In marshes and sloughs and along borders of ponds and streams. Along the Atlantic and Gulf Coasts.

Map 311. *Proserpinaca pectinata.*   Map 312. *Proserpinaca palustris.*

REFERENCES

Fernald, M. L. Two new Myriophyllums and a species new to the United States. Rhodora 21:120–124. 1919.
Fassett, N. C. *Myriophyllum* (new varieties). Rhodora 41:524–525. 1939.

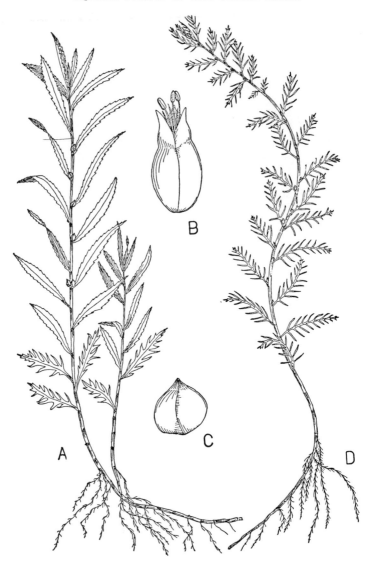

Fig. 128. *Proserpinaca palustris* (A–C) and *P. pectinata* (D). (A) Plant; x 1/2. (B) Flower; x 10. (C) Achene; x 4. (D) Plant; x 1/2.

UMBELLIFERAE: Parsley Family

Aromatic herbs, mostly with hollow, jointed stems and alternate leaves, the petioles expanded or sheathing at base. Flowers small, perfect, in simple or compound umbels or rarely heads. Calyx entire or 5-toothed, fused with the ovary; petals and stamens 5, attached on a disk on the ovary; pistil 2-celled, with 2 styles which are enlarged at the base into the stylopodium. Fruit 2 dry, seedlike carpels (mericarps) splitting when mature. Mericarps 1-seeded, with 5 longitudinal ridges (primary ribs) and sometimes 4 intermediate ribs in the intervals between the primary ribs, oil tubes or oil-secreting tissues in the pericarp. — A large family with aquatic members in several genera.

KEY TO GENERA

1. Leaves simple.
   2. Leaves reduced to hollow, cylindrical petioles; flowers in umbels.
      3. Plants with creeping stems .... *Lilaeopsis*
      3. Plants erect .... *Oxypolis*
   2. Leaves with expanded blades.
      4. Blade linear to lanceolate, margin with spiny teeth; flowers in heads; stem erect .... *Eryngium*
      4. Blade peltate or cordate, palmately veined; flowers in umbels or various clusters; stems creeping .... *Hydrocotyle*
1. Leaves compound; flowers in an umbel.
   5. Fruit with 1 oil tube between each pair of ridges; plant stoloniferous.
      6. Leaves once-pinnately compound .... *Oxypolis*
      6. Leaves decompound .... *Oenanthe*
   5. Fruit with 2 or more oil tubes between each pair of ridges.
      7. Fruit globose or nearly so, with corky pericarp and inconspicuous ridges; plant with rhizomes .... *Berula*
      7. Fruit ovate to oblong, with prominent corky ridges; plant without rhizomes .... *Sium*

BERULA

Aquatic perennials with erect or creeping stems rooting at the nodes. Leaves 1-pinnate, the 5 to 9 pairs of leaflets variable, crenate, serrate, or laciniate. Flowers white, in compound umbels with narrow involucral bracts. Carpels nearly globose, with depressed stylopodium and with slender ribs and thick corky pericarp; oil tubes numerous. — A few species widespread in temperate regions of both hemispheres.

Berula pusilla (Nutt.) Fernald (*B. erecta* [Huds.] Cov.). Fig. 132, C. Map 313.

Local in shallow ponds, swamps, swales, and streams; sometimes forming extensive patches submersed in clear, spring-fed brooks. Widespread.

Map 313. *Berula pusilla.*   Map 314. *Eryngium aquaticum.*

ERYNGIUM: Eryngo

Perennials with tough, linear to lanceolate leaves with entire, toothed, or spiny margin. Flowers in axils of blue or white bracts, sessile, in dense bracteate heads. Fruit ovoid. — About 200 species widely distributed in temperate and subtropical regions; mostly terrestrial.

KEY TO SPECIES OF ERYNGIUM

1. Stems erect; leaves spiny .... *E. aquaticum*
1. Stems prostrate, rooting at the nodes; leaves not spiny .... *E. prostratum*

Eryngium aquaticum L. Fig. 131, A. Map 314.

Borders of ponds, streams, and ditches. Mostly near the coast; from New Jersey to Florida and Texas.

Eryngium prostratum Nutt.

In sloughs and along streams. Missouri to Florida and Texas.

HYDROCOTYLE: Water Pennywort

Low perennial herbs with slender, creeping stems, sometimes bearing small tubers. Leaves orbicular-peltate or reniform, on long, erect petioles direct from the rootstock or on leafy, branching stems; blades floating or emersed. Flowers small, white, in simple or proliferating umbels without an involucre. Fruit nearly orbicular, laterally compressed. Mericarps with 5 primary ribs and sometimes also 4 narrow secondary ribs. — About 75 species widely distributed in both hemispheres, many of them in water or wet places.

KEY TO SPECIES OF HYDROCOTYLE

1. Leaves peltate; peduncles as long as petioles, but from slender, creeping rootstocks.
    2. Umbels simple .... *H. umbellata*
    2. Umbels proliferous.
        3. Umbels forming an interrupted spike; pedicels short or none . *H. bonarensis*
        3. Umbels branching; pedicels slender, up to 2 cm. long .... *H. verticillata*
1. Leaves not peltate.
    4. Umbels sessile or nearly so, axillary; leaves thin, short-petioled .. *H. americana*
    4. Umbels peduncled; leaves thick, with elongated petioles ... *H. ranunculoides*

Hydrocotyle americana L. Fig. 129, B. Map 315.

On margins of ponds and streams and in marshes. Chiefly in the northeastern states and the Middle West.

Hydrocotyle bonarensis Lam. Fig. 130, C–D. Map 316.

In marshes and on low wet ground. Southeastern states; naturalized in many places.

Map 315. *Hydrocotyle americana.*

Map 316. *Hydrocotyle bonarensis.*

Map 317. *Hydrocotyle ranunculoides.*

Map 318. *Hydrocotyle verticillata.*

Hydrocotyle ranunculoides L.f. Fig. 130, A–B. Map 317.

In ponds and lakes and on margins of streams; mostly on sandy bottom. Atlantic Coast; local in the South and westward.

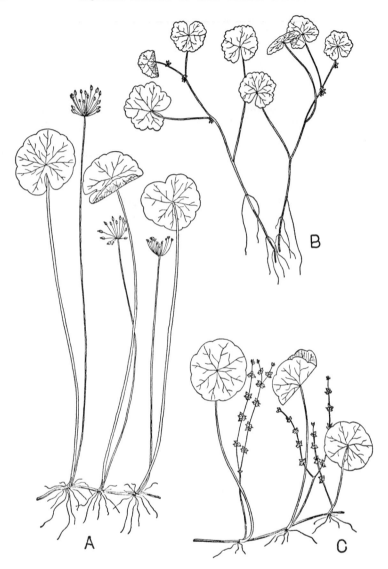

Fig. 129. *Hydrocotyle umbellata* (A), *H. americana* (B), and *H. verticillata* (C). All 1/2 natural size.

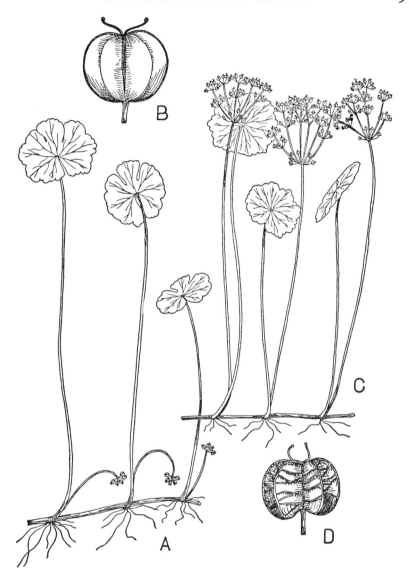

Fig. 130. *Hydrocotyle ranunculoides* (A, B) and *H. bonarensis* (C, D).
(A) Plant; x 1/2. (B) Fruit; x 10.
(C) Plant; x 1/2. (D) Fruit; x 10.

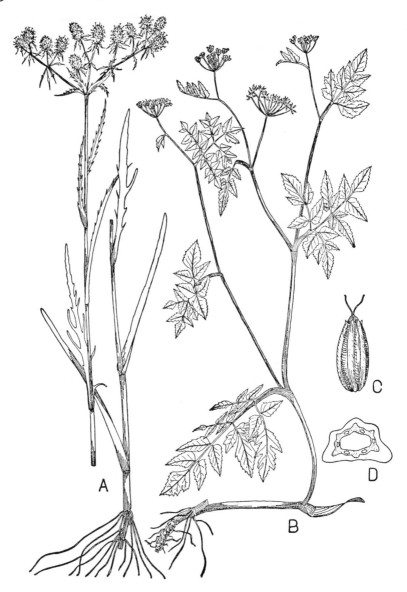

Fig. 131. *Eryngium aquaticum* (A) and *Oenanthe sarmentosa* (B–D).
(A) Plant; x 1/2.
(B) Plant; x 1/2. (C) Fruit; x 10. (D) Cross-section diagram of a "seed" (mericarp).

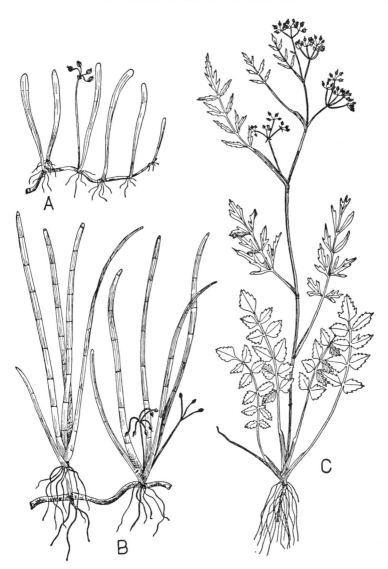

Fig. 132. *Lilaeopsis lineata* (A), *L. occidentalis* (B), and *Berula pusilla* (C), All 1/2 natural size.

Hydrocotyle umbellata L. Fig. 129, A. Map 319.

In shallow ponds and along borders of lakes and slow streams. Common in the northeastern states; local in the central states.

Hydrocotyle verticillata (A. Rich.) Fernald. Fig. 129, C. Map 318.

In marshes and shallow ponds and along streams. Chiefly along the Atlantic and Gulf Coasts.

MAP 319. *Hydrocotyle umbellata.*

MAP 320. *Lilaeopsis carolinensis.*

LILAEOPSIS

Low perennials with creeping rootstocks and erect, simple leaves (reduced to hollow, cylindrical, and subulate or flattened petioles jointed by thin transverse partitions). Flowers white or pale-yellow, in small, simple umbels with minute involucral bracts. Fruit globose, slightly compressed laterally, glabrous. Mericarps with all or with only the lateral ribs corky; oil tubes solitary in the intervals. — About 6 species, of wide distribution.

KEY TO SPECIES OF LILAEOPSIS

1. Peduncles longer than the leaves; lateral ribs on fruit corky-thickened, more prominent than the dorsal, noncorky ribs .... *L. lineata*
1. Peduncles shorter than the leaves.
    2. All ribs on fruit corky-thickened, the laterals more prominent .. *L. carolinensis*
    2. Only the lateral ribs corky-thickened .... *L. occidentalis*

Lilaeopsis carolinensis Coult. and Rose. Map 320.

Near the coast. From Virginia to South Carolina and Florida.

Lilaeopsis lineata (Michx.) Greene (*L. chinensis* Michx.). Fig. 132, A. Map 322.

Local in brackish marshes near the coast and on tidal flats along the banks of rivers. Maine to Florida.

Lilaeopsis occidentalis Coult. and Rose. Fig. 132, B. Map 321.

In tidal marshes, along banks of streams near the coast, rarely along shores of lakes. Along the Pacific Coast.

MAP 321. *Lilaeopsis occidentalis.*

MAP 322. *Lilaeopsis lineata.*

OENANTHE: WATER CELERY

Aquatic perennials with succulent, glabrous, often creeping stems and decompound or pinnate leaves; leaflets ovate, acuminate, serrate, often lobed at base. Umbels compound, usually with an involucre of a few linear bracts; involucels with numerous bractlets. Flowers white. Fruit globose or nearly so; mericarps with rounded, corky ridges; oil tubes solitary in the intervals. — About 30 species, mostly of the Eastern Hemisphere.

Oenanthe sarmentosa Presl. Fig. 131, B–D. Map 323.

Common in ponds, sloughs, intermittent pools, and ditches on the lowlands along the Pacific Coast from western Washington to northern California. The stems root at the nodes, branch, and form a dense cover over large areas.

MAP 323. *Oenanthe sarmentosa.*

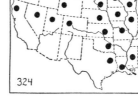

MAP 324. *Sium suave.*

OXYPOLIS: DROPWORT

Erect, glabrous, aquatic herbs with hollow stems. Leaves 1-pinnate or reduced to cylindrical, hollow petioles. Flowers white, in compound umbels with involucral bracts. Fruit ovoid to obovoid, with short, conical stylopodium, flattened parallel to the line of division; ribs narrow; oil tubes solitary in the intervals and 2 to 6 on the flattened side.

KEY TO SPECIES OF OXYPOLIS

1. Leaves reduced to cylindrical, hollow, nodose petioles .... *O. canbyi*
1. Leaves 1-pinnate, with 3 to 9 leaflets .... *O. rigidior*

Oxypolis canbyi (Coult. and Rose) Fernald. Map 325.

In marshes and along margins of ponds and streams. Mostly along the South Atlantic Coast.

Oxypolis rigidior (L.) Coult. and Rose. Map 326.

In ponds and marshes and along streams. Widespread in the eastern United States.

MAP 325. *Oxypolis canbyi*.

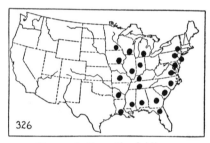

MAP 326. *Oxypolis rigidior*.

SIUM: WATER PARSNIP

Erect, glabrous perennials with hollow stems and 1-pinnate leaves, the leaflets serrate or, in submersed forms, finely dissected or laciniately lobed. Flowers white, in compound umbels with slender involucral bracts. Fruit ovoid or oblong, with prominent, corky ribs and 1 to 3 oil tubes in each interval; stylopodium depressed.

Sium suave Walt. Fig. 133. Map 324.

Shallow bays of lakes, ponds, slow streams, and marshes. Widespread except in the Southwest.

REFERENCES

Coulter, J. M., and J. N. Rose. Monograph of the North American Umbelliferae. Contr. U. S. Nat. Herb. 7:1–256. 1900.

Fernald, M. L. *Oxypolis canbyi*. Rhodora 41:139. 1939.

———. *Berula pusilla*. Rhodora 44:189–191. 1942.

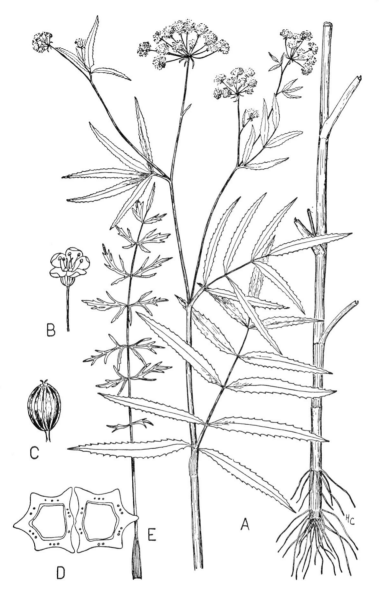

FIG. 133. *Sium suave.*
(A) Plant; x 1/3. (B) Flower; x 4. (C) Fruit; x 4. (D) Diagram of cross section of fruit; x 8. (E) Leaf from submersed plant; x 1/3. (From Muenscher, 1939.)

PRIMULACEAE: Primrose Family

Herbs with simple or, rarely, divided leaves. Flowers perfect, regular, axillary or in clusters. Petals 4 to 8, separate, fused, or wanting. Stamens as many as the petals and opposite them. Pistil with single style and superior, 1-celled ovary with free-central placenta. Fruit a many-seeded capsule. Seeds with endosperm. — Mostly terrestrial plants widely distributed in the Northern Hemisphere.

KEY TO GENERA

1. Leaves finely dissected, mostly submersed; plant often floating ....*Hottonia*
1. Leaves simple, with entire margin, mostly opposite.
    2. Leaves not dotted, sessile; flowers pink, solitary in the leaf axils .... *Glaux*
    2. Leaves dotted, petioled, often with bulblets in their axils; flowers yellow, in a terminal raceme ....,........................*Lysimachia*

GLAUX: Sea Milkwort

Low, erect or ascending perennials from creeping rootstocks, with opposite, sessile, entire, oblong to linear leaves. Flowers solitary, axillary, nearly sessile, white or pink to crimson. Calyx bell-shaped, 5-cleft, with broad petal-like lobes; petals wanting. Stamens 5, attached to the base of the calyx and alternating with the lobes. Capsule 5-valved, few-seeded. — Two species; only the following known from the United States.

Glaux maritima L. Fig. 134, A–C. Map 327.

Local in salt marshes, brackish bays, and tidal mud flats. North Atlantic and Pacific Coasts; local inland in the western states, especially in the Great Basin.

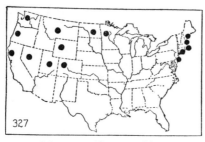

Map 327. *Glaux maritima.*

Map 328. *Hottonia inflata.*

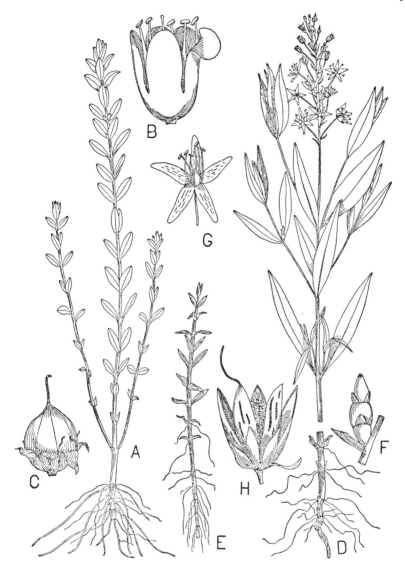

Fig. 134. *Glaux maritima* (A–C) and *Lysimachia terrestris* (D–H).

(A) Plant; x 1/2. (B) Flower; x 5. (C) Capsule; x 5.
(D) Plant, terrestrial form; x 1/2. (E) Plant, submersed form; x 2. (F) "Bulblet" from leaf axil; x 4. (G) Flower; x 2. (H) Dehiscing capsule with style remains and 5 valves; x 5.

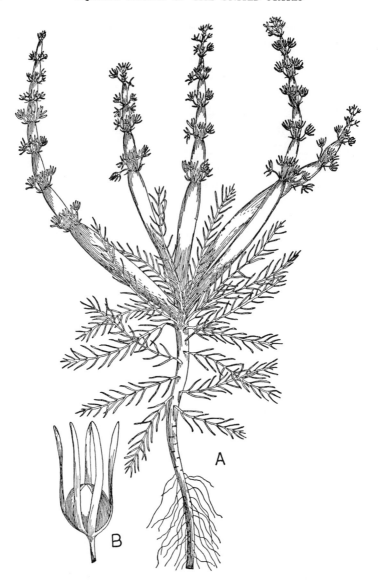

Fig. 135. *Hottonia inflata*.
(A) Plant; x 1/2. (B) Capsule; x 5.

HOTTONIA: Featherfoil, Water Violet

Perennials with floating and rooting stems and alternate, finely dissected, submersed leaves. Flowers in whorls on clustered, erect, nearly leafless, hollow or inflated stems. Calyx with 5 linear divisions; corolla with short tube and 5-parted limb; stamens 5, included. Capsule 5-valved, the valves cohering at base and summit. — Two species, only the following known from the United States.

Hottonia inflata Ell. Fig. 135. Map 328.

In pools and ditches. Chiefly in the southern states and in the Mississippi Valley; also locally introduced northward.

LYSIMACHIA: Loosestrife

Erect or creeping herbs with opposite, whorled or, rarely, some alternate leaves; blades commonly glandular-dotted. Flowers perfect, regular; calyx 5- to 6-parted; corolla rotate, deeply parted, the petals yellow with darker dots. Stamens with unequal filaments fused into a short tube at base. Capsule globose, several-seeded. — About 70 species, mostly terrestrial.

Lysimachia terrestris (L.) BSP. Fig. 134, D–H. Map 329.

Common in shallow ponds, pools, and marshes throughout the eastern United States. Locally introduced in cranberry bogs in the Pacific Northwest.

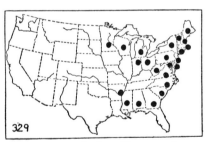 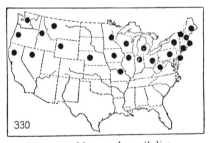

Map 329. *Lysimachia terrestris.*   Map 330. *Menyanthes trifoliata.*

MENYANTHACEAE: Buck-Bean Family

Perennial aquatic or bog plants with creeping stems. Leaves alternate, simple or trifoliate. Flowers perfect, regular, solitary or in clusters or racemes. Sepals 5, partly fused; corolla of 5 partly fused petals, rotate or funnelform. Stamens 5, with filaments partly fused with the corolla tube. Pistil of 2 fused carpels and short style. Fruit a capsule, dehiscent or indehiscent, several- to many-seeded. — A few genera, often treated as a subfamily under Gentianaceae.

KEY TO GENERA

1. Leaves simple, peltate or cordate, mostly floating; corolla nearly rotate ....*Nymphoides*
1. Leaves trifoliate, emersed; corolla funnelform ....*Menyanthes*

MENYANTHES: BUCK BEAN, BOG BEAN

Low herbs with thick, spongy, creeping rootstocks which have a very bitter taste. Leaves trifoliate, the petiole base usually persisting on the rootstock. Flowers white or pinkish in a raceme on a naked scape. Corolla funnelform, the upper surface white-bearded. Capsule dehiscent, with many hard, light-brown, lenticular seeds. — A single species, in both hemispheres.

Menyanthes trifoliata L. Fig. 136, A–C. Map 330.

Common in shallow ponds and lakes, on floating moors, and in acid bogs. Widespread across the northern United States except on the Great Plains.

NOTE. The seed coats of the buck bean are very durable. They are frequently found in large numbers in acid peat bogs at depths of several meters below the surface.

NYMPHOIDES: FLOATING HEART

Submersed plants with floating leaves. Leaves all basal with long petioles or some from branched, leafy stems; blades with a deep basal sinus, cordate, orbicular, or reniform, the margin usually crenate or undulate. Many of the petioles bear a bud and a cluster of fleshy, spurlike roots near the upper part. Flowers in small clusters on the petioles, often with the root clusters, or solitary on axillary peduncles on leafy stems. Corolla yellow or white, nearly rotate, each lobe with a basal, glandular appendage. Seeds with glossy, hard covering. — About 20 species of temperate and tropical regions.

KEY TO SPECIES OF NYMPHOIDES

1. Petioles without clusters of roots; leaves mostly from branching stems; flowers yellow, axillary; seeds flat, with a fringe of glands along the margin .. *N. peltatum*
1. Petioles with clusters of roots; leaves mostly basal, with slender petioles; flowers white, in clusters on petioles.
    2. Leaf blades ovate to ovate-orbicular, mostly less than 5 cm. long, not darkpunctate beneath; seeds smooth ....*N. cordatum*
    2. Leaf blades orbicular to reniform, mostly 8 to 15 cm. in diameter, darkpunctate or pitted beneath; seeds glandular, warty ....*N. aquaticum*

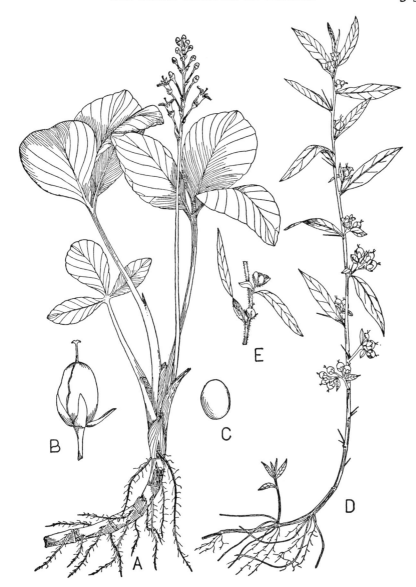

Fig. 136. *Menyanthes trifoliata* (A–C), *Hydrolea affinis* (D), and *H. quadrivalvis* (E).
(A) Plant; x 1/3. (B) Capsule; x 3. (C) Seed; x 5.
(D) Plant; x 1/2.
(E) Portion of plant; x 1/2.

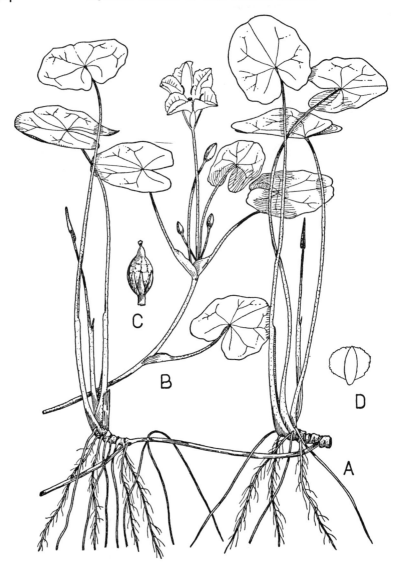

Fig. 137. *Nymphoides peltatum*.

(A) Sketch of small plant; x 1/3. (B) Flowering branch; x 1/3. (C) Mature capsule; x 1/2. (D) Petal; x 1/3.

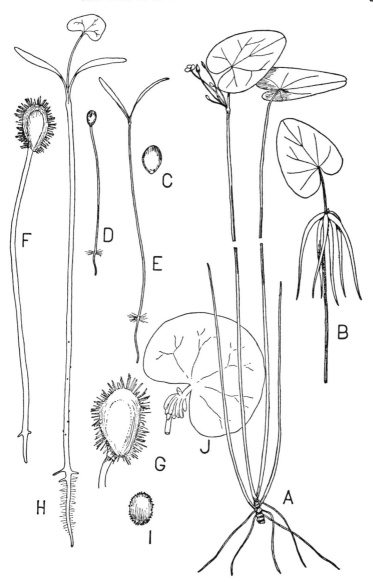

FIG. 138. *Nymphoides cordatum* (A-E), *N. peltatum* (F-H), and *N. aquaticum* (I, J).

(A) Portion of flowering plant; x 1/2. (B) Leaf with cluster of roots and bud on petiole; x 1/2. (C) Seed; x 6. (D) Germinating seed; x 3. (E) Seedling; x 3.

(F) Germinating seed; x 3. (G) Seedling protruding from seed coat; x 5. (H) Seedling, 2 weeks after seed germination; x 3.

(I) Seed with short papillae; x 8. (J) Leaf with bud and roots on petiole; x 1/2.

Nymphoides aquaticum (Walt.) Fernald. Fig. 138, I–J. Map 331.

In ponds and slow streams. In the southern states; locally introduced northward.

Nymphoides cordatum (Ell.) Fernald. Fig. 138, A–E. Map 332.

Common in lakes and ponds; mostly in acid waters with sandy bottom. Eastern and southern states and local in the Great Lakes region.

MAP 331. *Nymphoides aquaticum.*   MAP 332. *Nymphoides cordatum.*

Nymphoides peltatum (Gmel.) Brit. and Rendle. Figs. 137; 138, F–H. Map 333.

In rivers, slow streams, and ponds. Introduced from Eurasia. Naturalized in the Hudson River and escaped in widely separated places.

MAP 333. *Nymphoides peltatum.*

REFERENCE

Fernald, M. L. *Nymphoides cordatum.* Rhodora 40:338–340. 1938.

HYDROPHYLLACEAE: WATERLEAF FAMILY

Herbs with alternate or, rarely, opposite, rather fleshy leaves. Flowers perfect, regular, mostly in 1-sided or coiled clusters or, rarely, in axillary clusters. Sepals 5, fused; petals 5, fused into a tubular or nearly rotate corolla, mostly blue or white; stamens 5, inserted on the corolla;

carpels 2, fused into a compound pistil. Ovary superior, 1-celled, with 2 parietal placentae, or 2-celled; styles 2 or 2-cleft. Fruit a 2-valved capsule with 4 to many seeds. Seeds pitted or reticulate on the surface, with large endosperm. — A small group; most of its members are terrestrial.

HYDROLEA

Herbaceous perennials with alternate, entire leaves, sometimes with axillary spines. Corolla blue, 5-cleft, almost rotate or short-campanulate. Ovary 2-celled, with large, fleshy placentae projecting into the cells. Capsule globular, 2- or 4-valved, many-seeded. Seeds minute, ribbed. — Several species, mostly of warm or tropical regions; a few aquatics.

KEY TO SPECIES OF HYDROLEA

1. Calyx lobes as long as corolla; plant glabrous throughout . . . . *H. affinis*
1. Calyx lobes shorter than corolla; plant hairy above . . . . *H. quadrivalvis*

Hydrolea affinis Gray. Fig. 136, D. Map 334.
   In shallow ponds and swamps and on banks of streams. Lower Mississippi Valley.

Hydrolea quadrivalvis Walt. Fig. 136, E. Map 335.
   Along banks of streams and in low swampy places. Atlantic Coast states from Virginia to Florida.

MAP 334. *Hydrolea affinis.*

MAP 335. *Hydrolea quadrivalvis.*

SCROPHULARIACEAE: Figwort Family

Chiefly herbs with opposite or alternate leaves. Flowers perfect; corolla gamopetalous, irregular or regular; calyx 3- to 5-lobed, cleft, or toothed, often subtended by 2 sepal-like bracts; stamens 4 (sometimes 2 or 5), usually 2 long and 2 short, inserted on the corolla tube; carpels

2, fused, forming a 2-celled capsule with axile placenta and numerous seeds with endosperm. — A large family composed mostly of terrestrial plants, but several genera contain aquatic species.

KEY TO GENERA

1. Plants without leafy stems; leaves in a small rosette .... *Limosella*
1. Plants with leafy stems, erect, creeping, or floating.
    2. Sepals fused at least ½ their length.
        3. Corolla 20 to 40 mm. long, yellow .... *Mimulus*
        3. Corolla 2 to 15 mm. long.
            4. Corolla less than 3 mm. long, whitish .... *Hemianthus*
            4. Corolla more than 5 mm. long, violet-blue .... *Lindernia*
    2. Sepals distinct except at base.
        5. Flowers in axillary racemes, blue; (stamens 2) .... *Veronica*
        5. Flowers in the leaf axils, or in a simple, terminal raceme.
            6. Corolla irregular, with lobes shorter than the tube.
                7. Corolla yellow or white .... *Gratiola*
                7. Corolla violet-blue .... *Lindernia*
            6. Corolla nearly regular, bell-shaped, with lobes about as long as the tube.
                8. Pedicels with two bractlets; styles united.
                    9. Leaves with 1 main vein; corolla white; plant not scented .. *Bramia*
                    9. Leaves with 5 to 7 main veins; corolla blue; plant lemon-scented
                    .... *Hydrotrida*
                8. Pedicels without bractlets; corolla white; leaves entire or nearly so; styles 2, distinct.
                    10. Stamens 4 .... *Macuillamia*
                    10. Stamens 2 .... *Herpestis*

BRAMIA

Small perennial herbs with creeping or ascending stems and opposite, entire, sessile leaves with 1 main vein. Flowers on slender, axillary pedicels with 2 bractlets. Calyx of 5 distinct sepals, the upper largest, ovate; corolla nearly regular, bell-shaped, white, of 5 unequal lobes; stamens 4, nearly equal. Styles united; capsule turgid, many-seeded. — The only known species reaches its greatest development in the tropical regions of both hemispheres.

Bramia monnieri (L.) Pennell. Fig. 139, A–B. Map 336.

In fresh- and brackish-water pools and along their sandy shores. Chiefly on the Coastal Plain but also farther inland.

HERPESTIS

Small plants with creeping or lax stems; leaves opposite, nearly round, palmately veined, entire, somewhat clasping. Flowers on slender pedicels without bractlets; calyx with 5 distinct sepals, the upper cordate; corolla bell-shaped, white; stamens 2; pistil with 2 distinct styles. — A single species.

MAP 336. *Bramia monnieri*.

Herpestis rotundifolia C. F. Gaertn. Fig. 139, C–D. Map 337.

Very local in shallow ponds and streams and on muddy shores on the Coastal Plain from Maryland to Florida.

HYDROTRIDA

Slender, lemon-scented aquatics with rhizomes and emersed or floating stems; leaves opposite, with clasping, crenate blades with glandular dots and 5 to 7 veins. Flowers on short, axillary pedicels with 2 bractlets just below the calyx; calyx with 5 sepals, the upper cordate; corolla bell-shaped, nearly regular, blue, pubescent within; stamens 4; pistil with styles united; ovary surrounded by bristles at the base; capsule turgid, acute, many-seeded.

Hydrotrida caroliniana (Walt.) Small. Fig. 139, H. Map 338.

Common in shallow ponds and slow streams and along their sandy shores. Chiefly in the southeastern states.

MAP 337. *Herpestis rotundifolia*.

MAP 338. *Hydrotrida caroliniana*.

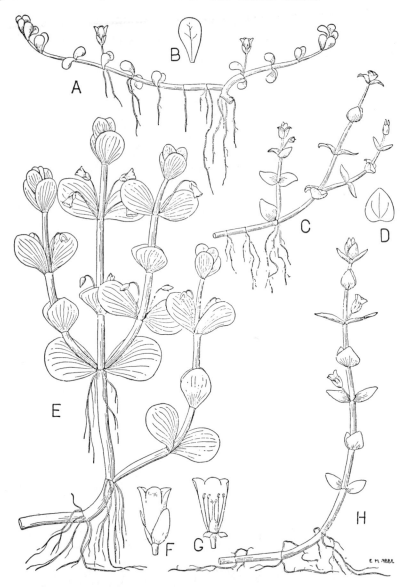

Fig. 139. *Bramia monnieri* (A, B), *Herpestis rotundifolia* (C, D), *Macuillamia rotundifolia* (E–G), and *Hydrotrida caroliniana* (H).

(A) Plant; x 1/2. (B) Leaf; x 1 1/2.
(C) Plant; x 1/2. (D) Leaf; x 1.
(E) Plant; x 1/2. (F) Flower; x 2. (G) Vertical section of flower; x 2.
(H) Plant; x 1/2.

MACUILLAMIA: Water Hyssop

Slender herbs with creeping stems and ascending branches, submersed below and with floating tufts of leaves. Leaves opposite, entire, nearly round, palmately veined, glandular-dotted. Flowers several in a leaf axil, each on a slender pedicel without bractlets; calyx with upper sepal oblong or nearly round; corolla campanulate, white; stamens 4; pistil with 2 distinct styles; capsule obtuse. — A few species widespread in the Western Hemisphere.

KEY TO SPECIES OF MACUILLAMIA

1. Outer sepal nearly orbicular, strongly veined; pedicels 10 to 15 mm. long .... M. rotundifolia
1. Outer sepal oblong, without prominent veins; pedicels 5 to 7 mm. long . M. repens

Macuillamia repens (Swartz) Pennell. Map 340.

Very local in shallow ponds along the South Atlantic Coast. Possibly introduced from tropical America where it is common.

Macuillamia rotundifolia (Michx.) Raf. Fig. 139, E–G. Map 339.

In shallow ponds with muddy bottom. In the Mississippi Valley and the Great Plains states.

Map 339. *Macuillamia rotundifolia.* Map 340. *Macuillamia repens.*

GRATIOLA: Hedge Hyssop

Low annual or perennial herbs with opposite, mostly serrate or crenate, glandular or clammy leaves. Flowers on axillary pedicels, usually with 2 bractlets as long as and close to the 5-parted calyx; corolla 2-lipped, the upper entire or 2-cleft, the lower 3-cleft, yellow, white, or pale-purplish, with a beard of stiff hairs inside; stamens 2 fertile and usually 2 sterile; anthers with a broad connective; capsule usually

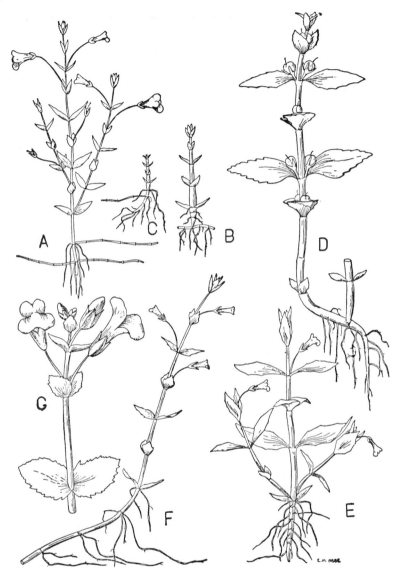

Fig. 140. *Gratiola lutea* (A–C), *G. virginiana* (D), *G. neglecta* (E), *G. viscidula* (F), and *Mimulus guttatus* (G).

(A) Habit of flowering plant with slender rhizomes. (B) Sterile plant from shallow water. (C) Sterile plant from deep water.
(D–G) Plants showing habit. All x 1/2.

4-valved, many-seeded. — About 20 species of temperate regions and on mountains in tropical regions.

KEY TO SPECIES OF GRATIOLA

1. Plants perennial from slender rhizomes; sepals much longer than the capsule; stem leaves clasping.
   2. Corolla yellow .... *G. lutea*
   2. Corolla white or purplish, with purple lines in throat .... *G. viscidula*
1. Plants annual, without rhizomes; sepals about as long as the capsule; stem leaves not clasping.
   3. Pedicels slender; corolla greenish-yellow, the throat with filiform hairs on the upper side .... *G. neglecta*
   3. Pedicels stout; corolla white, the throat with purple lines, with clavate hairs on the upper side .... *G. virginiana*

Gratiola lutea Raf. (*G. aurea* Muhl.). Fig. 140, A–C. Map 341.

Common in shallow water and on shores, both sandy and peaty, of lakes, ponds, and streams. Chiefly in the northeastern states.

NOTE. Small, sterile plants sometimes form turflike beds over extensive areas of sandy bottom in clear water several meters deep in Connecticut, New Hampshire, New York, and Wisconsin. Such plants have been named *G. lutea* f. *pusilla* Fassett. They represent only a deep-water phase or state.

Gratiola neglecta Torr. Fig. 140, E. Map 342.

In shallow pools early becoming desiccated. Widespread across the United States.

Gratiola virginiana L. Fig. 140, D. Map 343.

Widespread and locally common in wet depressions and on muddy banks of ponds and streams, chiefly between high- and low-water levels.

Gratiola viscidula Pennell. Fig. 140, F. Map 344.

In swales and along streams. Chiefly on the Piedmont from Delaware to Georgia.

HEMIANTHUS

Dwarf, tufted or creeping annuals 1 to 10 cm. high; leaves sessile, rounded or spatulate, entire. Flowers minute, on slender, axillary, recurving pedicels; calyx 4-toothed; corolla white or purplish, remaining closed; stamens 2 fertile and 2 sterile; capsule globular, several-seeded. — About 10 species, mainly West Indian.

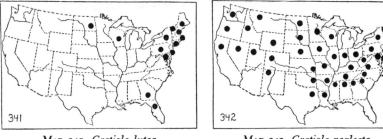

MAP 341. *Gratiola lutea*.   MAP 342. *Gratiola neglecta*.

MAP 343. *Gratiola virginiana*.   MAP 344. *Gratiola viscidula*.

Hemianthus micranthemoides Nutt. Fig. 141, C. Map 345.

Local on gravelly or silty shores of rivers, in fresh water between tides, from the Hudson River to Virginia. In the Hudson River locality

MAP 345. *Hemianthus micranthemoides*.

the plants are covered to a depth of about 1 meter when the tide is high and at low tide they are kept moist by water seeping from springs.

LIMOSELLA: MUDWORT

Small annuals with creeping stems forming mats only a few centimeters high. Leaves in clusters or alternate, fleshy, subulate or with elliptic to ovate blades, 2 to 4 cm. long. Flowers small, solitary on

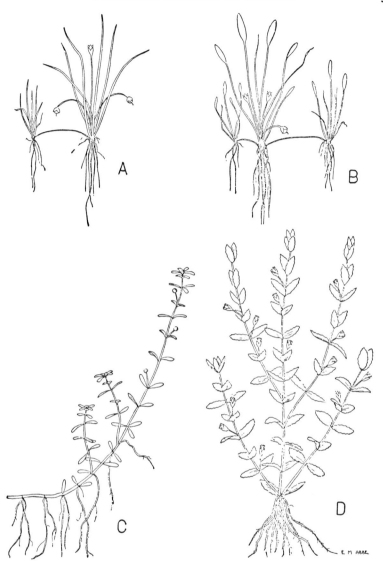

Fig. 141. *Limosella subulata* (A), *L. aquatica* (B), *Hemianthus micranthemoides* (C), and *Lindernia dubia* (D). All x 1, except D, x 1/2.

axillary, naked peduncles; calyx 5-toothed; corolla bell-shaped, 5-cleft, white or purple; stamens 4; capsule globular, many-seeded. — About 15 species of wide distribution; mostly aquatics.

KEY TO SPECIES OF LIMOSELLA

1. Leaves with elliptic or ovate blades; style up to 0.4 mm. long ....*L. aquatica*
1. Leaves with blades no wider than petiole, subulate; style 1 mm. long. *L. subulata*

Limosella aquatica L. Fig. 141, B. Map 347.

On muddy and sandy shores of fresh and brackish ponds and tidal shores of rivers. Chiefly near the North Atlantic Coast.

Limosella subulata Ives. Fig. 141, A. Map 346.

Muddy shores of ponds, lakes, and slow streams. In the western states.

Map 346. *Limosella subulata.*   Map 347. *Limosella aquatica.*

LINDERNIA: False Pimpernel

Small, branched or spreading, glabrous, leafy annuals, 1 to 3 dm. high. Leaves opposite (at least the lower), ovate to oblong, crenate. Flowers on axillary, naked peduncles or the upper appearing as in a raceme; calyx 5-lobed; corolla 5 to 10 mm. long, 2-lipped, the upper lip short, erect, 2-cleft, the lower larger and 3-cleft. Stamens 2 fertile and 2 sterile. Stigma 2-lobed; capsule ellipsoid to ovoid, many-seeded. — A large genus mostly of the Old World tropics. A few species are semiaquatic.

Lindernia dubia (L.) Barnhart. Fig. 141, D. Map 348.

Widely distributed along shores of ponds, lakes, and streams; common on muddy or sandy bottoms of desiccated ponds and ditches.

MIMULUS: Monkey Flower

Herbs with erect or suberect stems and opposite leaves. Lower leaves petioled, often with small, pinnate lobes, upper leaves oval, dentate. Flowers solitary or several on a peduncle in the axils of the upper leaves; calyx 5-angled, 5-toothed; corolla showy, yellow with orange spots, 2 to 4 cm. long, strongly 2-lipped, the upper lip erect or reflexed, 2-lobed, the lower spreading, 3-lobed; stigma 2-lobed; capsule turgid, 2-valved, many-seeded. — A large genus with many terrestrial species; the description is based on the following species.

MAP 348. *Lindernia dubia.*   MAP 349. *Mimulus guttatus.*

Mimulus guttatus Fisch. Monkey flower. Fig. 140, G. Map 349.

Common in slow streams, ditches, and spring-fed pools throughout the western states. Occasionally naturalized in the central and eastern states where it is sometimes grown under cultivation.

NOTE. In the Pacific Northwest this species, also associated with *Veronica americana,* often completely chokes open drainage ditches. It grows submersed during the winter and early spring and later produces emersed shoots with flowers.

VERONICA: Speedwell

This genus contains about 250 species, mostly terrestrial. The few aquatic species are perennial, stoloniferous or decumbent herbs rooting freely at the lower nodes. Leaves opposite, mostly serrate or crenate, sessile or petioled, somewhat fleshy. Flowers perfect, borne in opposite, axillary racemes; calyx mostly 4-parted; corolla wheel-shaped, nearly regular, blue or pale lavender; stamens 2; pistil with an entire style, single stigma, and flattened or turgid, 2-celled ovary; capsule about as long as wide, sometimes slightly notched, many-seeded.

KEY TO SPECIES OF VERONICA

1. Leaves, at least the upper, sessile and somewhat clasping; racemes with more than 25 flowers .... *V. anagallis-aquatica*
1. Leaves all with petioles; racemes mostly with fewer than 25 flowers.
    2. Leaf blades widest near the base, acute at apex .... *V. americana*
    2. Leaf blades widest at or above the middle, rounded at apex .... *V. beccabunga*

Veronica americana (Raf.) Schwein. American brooklime. Fig. 142, C–D. Map 350.

Widespread and common in brooks, ditches, and spring-fed ponds.

Veronica anagallis-aquatica L. Water speedwell. Fig. 142, A–B. Map 351.

Apparently naturalized from Eurasia. Widespread and locally abundant in brooks, ditches, and spring-fed ponds.

Map 350. *Veronica americana.*   Map 351. *Veronica anagallis-aquatica.*

Veronica beccabunga L. European brooklime. Fig. 142, E. Map 352.

Introduced from Eurasia. Very local in slow streams and in spring-fed ponds about fish hatcheries; also reported along muddy shores.

Map 352. *Veronica beccabunga.*

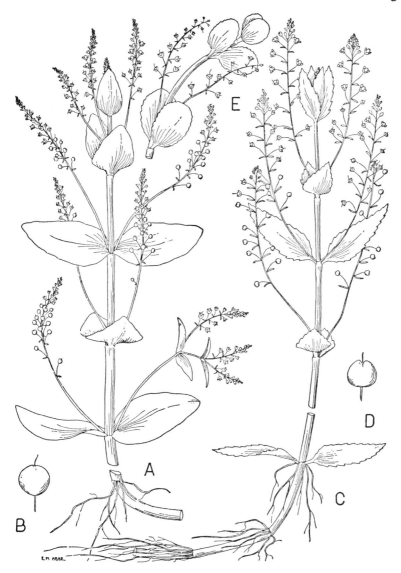

FIG. 142. *Veronica anagallis-aquatica* (A, B), *V. americana* (C, D), and *V. beccabunga* (E).

(A) Plant. (B) Capsule.
(C) Plant. (D) Capsule.
(E) Portion of plant.
All x 1/2, except B and D, x 3.

REFERENCES

Pennell, F. W. Scrophulariaceae of the West Gulf States. Proc. Acad. Nat. Sci. Phila. 73:459–535. 1921.

———. The Scrophulariaceae of eastern temperate North America. Acad. Nat. Sci. Phila. Monogr. 1:1–650. 1935.

LENTIBULARIACEAE: Bladderwort Family

Mostly delicate herbs with racemose or solitary, perfect flowers. Calyx and corolla 2-lipped; stamens 2; pistil with a 1-celled, superior ovary with free-central placenta. Fruit a capsule, with a few or many seeds without endosperm. — Widely distributed especially in the tropics.

UTRICULARIA: Bladderwort

Submersed or floating rootless aquatics with alternate or, rarely, whorled, flaccid, finely dissected or filiform, simple leaves. Many of the leaf segments of most species produce small bladders furnished with a little trap door and, usually, several minute bristles at the orifice. These bladders may trap small aquatic animals.

In the absence of true roots, even in the seedling stages, the branches and leaves may be modified in various ways to serve as anchoring or absorbing organs. In some species the stems are almost entirely subterranean, colorless, and rhizome-like. These produce erect, linear leaves or branches with a few reduced leaves. Many species produce leafy stems which are entirely free-floating or anchored only by the base. Such stems are sparsely branched, usually horizontal, and covered with finely dissected leaves, many of which are bladder-bearing. A few species produce two kinds of vegetative branches: the lower, imbedded in the mud, are colorless and produce numerous bladders or rootlike segments; the others, submersed or sometimes creeping on the surface of the bottom mud, are green. Flowers mostly on erect scapes or on short, axillary pedicels, in a few species in part cleistogamous.

Corolla yellow or purple, the upper lip erect or nearly so; the lower lip 3-lobed, spurred at the base; the palate of the lower lip often projecting so as nearly to close the throat, often bearded. Style short or wanting; stigma 1- or 2-lobed; capsule usually opening irregularly, rarely indehiscent. Seeds various and mostly characteristic for each species (Fig. 146). — About 250 species.

KEY TO SPECIES OF UTRICULARIA

1. Stems (except for the erect scape) subterranean, forming a mat of branched, slender rhizomes bearing slender, erect leaves or simple, erect branches with reduced or scalelike leaves and, rarely, a few small bladders.
   2. Bracts on the scape a single pair fused into a tube; flower solitary, purple ....*U. resupinata*
   2. Bracts on the scape several, alternate, not fused into a tube; scape mostly several-flowered.
      3. Pedicels slender, much longer than the bracts or flowers; corolla purple or yellow ....*U. subulata*
      3. Pedicels stout, shorter than the bracts or flowers; corolla yellow.
         4. Corolla much exceeding the calyx ....*U. cornuta*
         4. Corolla not exceeding the calyx ....*U. virgatula*
1. Stems floating, submersed or partly creeping on the soil surface, sometimes anchored at the base, the rest free, more or less leafy throughout with finely dissected leaves some or all of which bear bladders on some of the segments.
   5. Branches in whorls; flowers in racemes on scapes, purple ....*U. purpurea*
   5. Branches alternate, except the flowering scapes may have 1 whorl of inflated branches; flowers yellow.
      6. Scapes with a whorl of inflated branches below the raceme ......*U. inflata*
      6. Scapes without inflated branches.
         7. Leaf segments flat, with a midvein.
            8. Branches dimorphous, the lower almost leafless, mostly colorless, bearing bladders, the others green, leafy, almost bladderless; leaf segments with minutely spiny margin; pedicels erect in fruit ......*U. intermedia*
            8. Branches all similar, equally bladder-bearing and colored; leaf segments entire except near the tip; pedicels recurved in fruit ....*U. minor*
         7. Leaf segments filiform (not flattened), without a midvein.
            9. Plants minute, of delicate capillary branches; leaves reduced to a single segment, often bearing a bladder; bracts of the reduced scape sessile on the leafy stem, bearing 1 or, rarely, more pedicels not more than 1 cm. long; corolla less than 2 mm. long; capsule 1-seeded ..*U. olivacea*
            9. Plants larger; leaves mostly with several to many segments; flowers, at least some of them, in racemes on an elongated scape; capsule several-seeded.
               10. Stems floating or with some of the branches draping; leaves 2 to 5 times forked.
                  11. Scape stout, 6- to 21-flowered; leaf segments minutely serrate along the margin ....*U. vulgaris*
                  11. Scape slender, 1- to 5-flowered, usually also with a cleistogamous flower at its base attached to the leafy, horizontal stem; flowers often all cleistogamous, on short pedicels in the leaf axils; leaf segments not serrate on the margin ....*U. geminiscapa*
               10. Stems, at least in part, creeping on the bottom; scape slender.
                  12. Branches not radiating from base of scape; spur shorter than the lower lip of corolla; scape 1- to 3-flowered ....*U. gibba*

12. Branches radiating from base of scape; spur at least as long as the lower lip of corolla; scape 1- to 8-flowered.
   13. Spur tapering from base to tip; leaves all alike ..... *U. biflora*
   13. Spur cylindric above, conic at base; leaves not all alike, not all bladder-bearing .... *U. fibrosa*

Utricularia biflora Walt. Fig. 144, D. Map 353.

In shallow water of ponds and slow streams. Chiefly along the Atlantic Coast.

Utricularia cornuta Michx. Figs. 143, E–G; 146, G. Map 354.

Common on floating moors and also on wet, sandy shores. Throughout the eastern United States. Specimens with corollas reduced to less than 1 cm. broad have been named *U. juncea* Vahl.

Utricularia fibrosa Walt. Figs. 144, B–C; 146, B. Map 355.

In shallow pools and slow streams. Mostly in the pine-barren region of the Atlantic and Gulf Coast.

Utricularia geminiscapa Benj. Figs. 144, G–H; 145; 146, C. Map 356.

In ponds, ditches, slow streams, and bog holes; mostly in acid waters in peat bogs. Chiefly in the northeastern states.

Utricularia gibba L. Figs. 144, F; 146, H. Map 357.

Common in spring-fed pools and streams and in shallow ponds and depressions in wet boggy areas. Along the Atlantic Coast and in the Mississippi Valley.

Utricularia inflata Walt. Fig. 143, L. Map 358.

In ponds, pools, and ditches. Mostly on the Coastal Plain, from Maine to Florida. Smaller plants with shorter and fewer-flowered scapes have been called *U. radiata* Small or *U. inflata* var. *minor* Chapm.

MAP 353. *Utricularia biflora*.

MAP 354. *Utricularia cornuta*.

THE PLANTS ARRANGED BY FAMILIES 323

Map 355. *Utricularia fibrosa.*

Map 356. *Utricularia geminiscapa.*

Map 357. *Utricularia gibba.*

Map 358. *Utricularia inflata.*

Utricularia intermedia Hayne. Fig. 144, E. Map 359.
In shallow ponds and slow streams. Chiefly in the northeastern states.

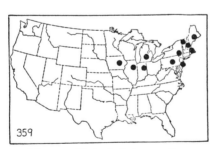

Map 359. *Utricularia intermedia.*

Utricularia minor L. Fig. 144, A. Map 360.
In shallow ponds and streams, usually on mucky bottom or shores. Mostly in the northeastern states and in the Great Lakes region.

Utricularia olivacea Wright. Fig. 143, K. Map 361.
Floating in shallow pools. Southern Florida.

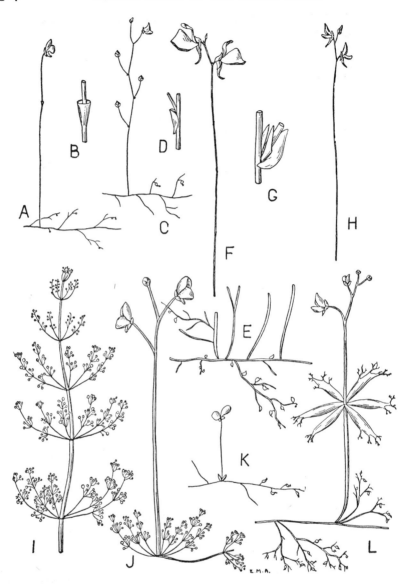

Fig. 143. *Utricularia resupinata* (A, B), *U. subulata* (C, D), *U. cornuta* (E–G), *U. virgatula* (H), *U. purpurea* (I, J), *U. olivacea* (K), and *U. inflata* (L). (A) Part of plant with flower; x 1/2. (B) Fused bracts of flower stalk; x 5. (C) Plant; x 1/2. (D) Peltate bract from base of pedicel; x 5. (E) Basal part of plant with erect, simple leaves and rootlike, branched leaves with bladders; x 2. (F) Flowers; x 1/2. (G) Bract and bractlets below flower; x 5. (H) Flowers; x 1/2. (I) Sterile shoot with whorled branches; x 1/2. (J) Erect flower-bearing shoot; x 1/2. (K) Plant showing flower with 2 bracts at base of scape; x 1. (L) Part of plant showing flowers subtended by whorl of inflated branches; x 1/2.

THE PLANTS ARRANGED BY FAMILIES 325

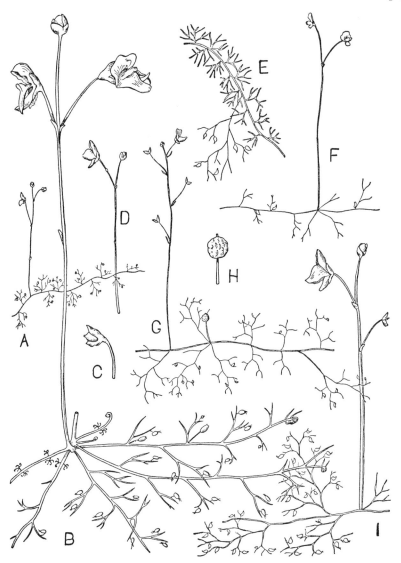

FIG. 144. *Utricularia minor* (A), *U. fibrosa* (B, C), *U. biflora* (D), *U. intermedia* (E), *U. gibba* (F), *U. geminiscapa* (G, H), and *U. vulgaris* (I). (A) Portion of plant with inflorescence; x 1/2. (B) Plant with inflorescence; x 2. (C) Young capsule; x 2. (D) Flowers; x 1/2. (E) Part of sterile plant; x 1/2. (F) Part of plant with inflorescence; x 1/2. (G) Portion of plant with inflorescence and solitary cleistogamous flower; x 1/2. (H) Capsule; x 3. (I) Portion of plant showing bladder-bearing leaves and inflorescence; x 1/2.

MAP 360. *Utricularia minor.*   MAP 361. *Utricularia olivacea.*

Utricularia purpurea Walt. Figs. 143, I–J; 146, D. Map 362.

Local in ponds and lakes, mostly in acid waters. Widespread along the Atlantic and Gulf Coasts and in the Great Lakes region.

MAP 362. *Utricularia purpurea.*

Utricularia resupinata Greene. Figs. 143, A–B; 146, F. Map 363.

On sandy bottom and shores of ponds and lakes. Atlantic Coast and Great Lakes region.

Utricularia subulata L. Figs. 143, C–D; 146, A. Map 364.

In springy places along streams and ponds. Mostly on the Coastal Plain, from Massachusetts to Florida. Forms with small, mostly cleistogamous, purplish flowers are by some authors segregated as a separate species, *U. cleistogama.*

Utricularia virgatula Barnh. Fig. 143, H. Map 365.

In shallow water and on wet, sandy shores of ponds and streams. Chiefly along the Atlantic Coast.

Utricularia vulgaris L. Figs. 144, I; 146, E. Map 366.

Widespread in ponds, lakes, and slow streams. Almost throughout the United States, except in the extreme South. This is a widespread species in Eurasia. Most of the American plants are more robust and

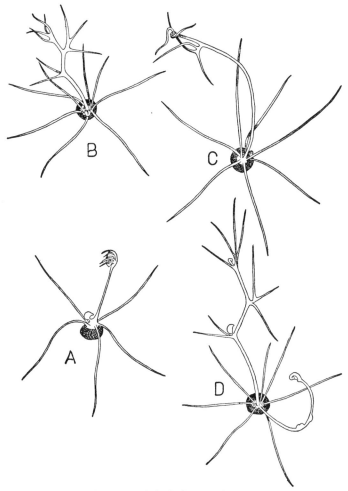

Fig. 145. *Utricularia geminiscapa.*

(A–D) Seedlings, 1 to 4 weeks after germination. Note the 5 to 6 cotyledons from a bulbous base and the absence of a primary root.

Map 363. *Utricularia resupinata.*   Map 364. *Utricularia subulata.*

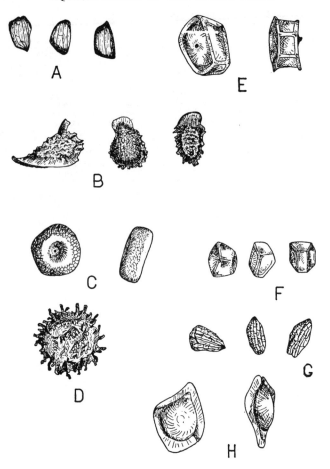

Fig. 146. Seeds of *Utricularia*.
(A) *Utricularia subulata*. (B) *U. fibrosa*. (C) *U. geminiscapa*. (D) *U. purpurea*. (E) *U. vulgaris*. (F) *U. resupinata*. (G) *U. cornuta*. (H) *U. gibba*. All about x 10.

Map 365. *Utricularia virgatula*.

Map 366. *Utricularia vulgaris*.

have a slenderer spur than the Eurasian plants. They are usually treated as a variety, *U. vulgaris americana* Gray.

REFERENCES

Rossbach, G. B. Aquatic Utricularias. Rhodora 41:113–128. 1939.
Lloyd, F. E. The traps of *Utricularia*. Proc. 6th Int. Bot. Congr. 1:54–73. 1935.
———. *Utricularia*. Biol. Rev. 10:72–110. 1935.
———. The structure and behavior of *Utricularia purpurea*. Canadian Jour. Res. 8:234–252. 1933.

ACANTHACEAE: Acanthus Family

Mostly herbs with opposite, simple leaves. Flowers perfect, irregular, in terminal or axillary, peduncled, bracteate spikes. Calyx 5-cleft; corolla mostly 2-lipped; stamens 2 or 4 (2 long and 2 short), inserted on the corolla tube. Pistil with filiform style, simple or 2-cleft stigma, and superior ovary. Fruit a 2-celled, loculicidal capsule, mostly flattened at right angles to the partition, few-seeded. Seeds flattened, without endosperm. — A large group, mostly of tropical regions.

JUSTICIA (*DIANTHERA*): Water Willow

Perennials with thick, creeping rootstocks and erect stems 4 to 10 dm. high with entire, opposite leaves. Flowers in short spikes on long, axillary peduncles. Calyx 5-parted. Corolla purple, with upper lip notched; the lower lip spreading and 3-parted. Stamens 2, with 2-celled, somewhat separated anthers. Capsule obovate, the base tapering into a stalk, flattened, 1- to 4-seeded. — Many species, mostly of tropical America.

Justicia americana (L.) Vahl. Fig. 147, F–I. Map 367.
Emersed in shallow lakes and streams or along wet shores subject to inundation. Eastern states westward to the Mississippi Valley.

PLANTAGINACEAE: Plantain Family

Chiefly low herbs with scapose stems and rosettes of simple, basal leaves. Flowers perfect or imperfect, in spikes or solitary, the pistillate flowers basal. Calyx 4-lobed; corolla tubular or rotate, or rarely 3-

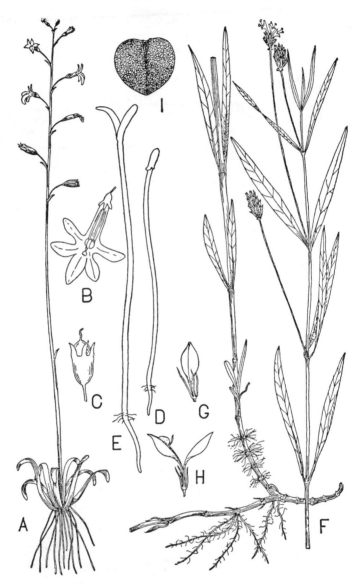

FIG. 147. *Lobelia dortmanna* (A–E) and *Justicia americana* (F–I).

(A) Plant; x 1/2. (B) Flower; x 1 1/2. (C) Capsule; x 1 1/2. (D) Seedling protruding from seed coat; x 5. (E) Seedling; x 5.

(F) Plant; x 1/2. (G) Capsule; x 1 1/2. (H) Capsule dehiscing; x 1 1/2. (I) Seed; x 5.

Map 367. *Justicia americana.*

Map 368. *Littorella americana.*

parted. Stamens 4 or 2, inserted on the corolla. Fruit a capsule or achene. — Three genera, about 225 species, mostly terrestrial.

KEY TO GENERA

1. Plants submersed, stoloniferous; scape with 1 staminate flower; pistillate flowers sessile; ovary 1-seeded .... *Littorella*
1. Plants emersed or in saline marshes; scape with many perfect flowers; ovary 2- to 4-seeded .... *Plantago*

LITTORELLA

Dwarf, submersed perennials with slender stolons and rosettes of awl-shaped leaves. Flowers small, imperfect. The staminate solitary on a slender, 1-bracteate or naked scape; calyx 4-parted and exceeding the narrow, 4-cleft corolla; stamens 4, exserted. The pistillate flowers 2 or 3 at the base of the scape, axillary in bracts; calyx of 3 or 4 unequal sepals; corolla cup-shaped, 3- to 4-toothed; pistil a 1-celled ovary with filiform style. Fruit an achene. — Four or 5 species, only the following known from the United States.

Littorella americana Fernald. Fig. 148, D–E. Map 368.

Infrequent in shallow ponds and lakes with silt-covered, gravelly bottom. A northern species of wide distribution, extending into the United States in New England, northern New York, and Minnesota.

PLANTAGO: Plantain

Leaves basal, prominently ribbed, linear, lanceolate to orbicular. Flowers small, greenish or white, in dense, bracted spikes terminating the scapes. Sepals 4, imbricated and usually persistent. Petals 4, fused, except for the lobes, into a tubular or rotate corolla, usually drying and clinging to the capsule. Stamens 4 or 2, with slender filaments. Fruit a 2-celled circumscissile capsule with 2 or several seeds. — About 200 species.

MAP 369. *Plantago maritima*.

Plantago maritima L. Seaside plantain. Fig. 148, A–C. Map 369.

Common in shallow, tidal pools in salt marshes and along seabeaches. North Atlantic Coast and North Pacific Coast. This rather variable complex species includes the linear-leaved, fleshy plantains along the seacoasts. These are sometimes separated into several species based upon the hairiness of the plants and the nature of the bracts, calyx, and capsule, characters that appear to be rather inconstant.

REFERENCES

Fernald, M. L. North American *Littorella*. Rhodora 20:61–62. 1918.
Muenscher, W. C. Occurrence of *Littorella americana* in New York. Rhodora 36:194. 1934.
Fernald, M. L. Maritime *Plantago* in North America. Rhodora 27:93–104. 1925.
Gregor, J. W. Experimental taxonomy IV. . . . Sea plantains allied to *Plantago maritima*. New Phytol. 38:293–332. 1939.

LOBELIACEAE: LOBELIA FAMILY

Plants with milky juice and alternate, simple leaves. Flowers perfect, irregular, arranged in racemes or spikes or axillary. Calyx tube fused with the ovary. Corolla 2-lipped, the lower lip 3-lobed, the upper 2-lobed. Stamens 5, usually the anthers and sometimes the filaments united into a tube. Pistil with a 2-celled or, rarely, 1-celled ovary and a single style with stigma usually fringed. Fruit a 1- or 2-celled capsule, opening near the top. Seeds with endosperm. — A large family, mostly terrestrial.

KEY TO GENERA

1. Corolla tube split down the upper side; perennials .... *Lobelia*
1. Corolla tube not split down the upper side; annuals .... *Howellia*

HOWELLIA

Delicate annuals with flaccid, sparingly branched stems and alternate, linear to subulate, entire leaves. Flowers axillary, the earlier

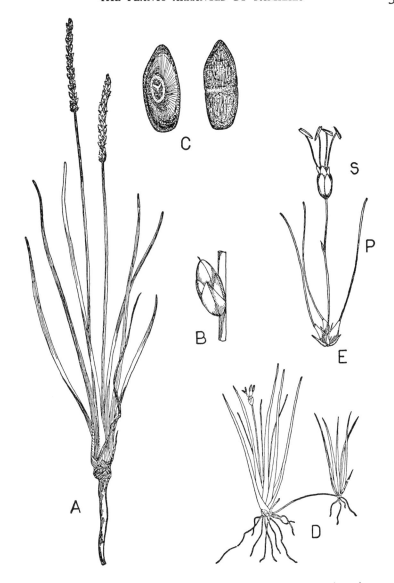

Fig. 148. *Plantago maritima* (A–C) and *Littorella americana* (D, E).
(A) Plant; x 1/2. (B) Capsule; x 5. (C) Seeds; x 10.
(D) Plant; x 1. (E) Inflorescence showing staminate (S) and pistillate (P) flowers; x 2.

sometimes cleistogamous. Calyx with short tube fused with the ovary and 5 narrow lobes. Corolla white or pale lavender, the tube short, cleft nearly to the base on one side, the lobes nearly equal. Anthers unequal, 2 short and 3 long. Capsule 1-celled, with 2 parietal placentae, few-seeded. — One species.

Howellia aquatica Gray. Map 370.

Local in shallow ponds. Oregon and Idaho.

MAP 370. *Howellia aquatica*.   MAP 371. *Lobelia dortmanna*.

LOBELIA

Perennial herbs with leaves mostly in basal rosettes, those on the stem often much reduced. Flowers with short calyx tube and 2-lipped, blue or, rarely, red corolla with the tube cleft to the base on what is apparently the upper side. Stamens 5, 2 of them with anthers bearded on top. Capsule many-seeded. — About 200 species, widely distributed, mostly terrestrial.

Lobelia dortmanna L. Water lobelia. Fig. 147, A–E. Map 371.

In shallow water and on wet shores, mostly on sandy bottom of acid ponds and lakes. Northeastern states, Great Lakes region, and the Pacific Northwest.

COMPOSITAE: Composite Family

Flowers small, clustered in a head on a receptacle surrounded by bracts forming an involucre. The surface of the receptacle may bear the flowers and small bracts (chaff) or it may be free of bracts, i.e., have a naked receptacle. The calyx is modified and reduced to awns, bristles, or scales (the pappus). The corolla, composed of 5 fused petals, may be tubular, as in the disk flowers, or strap-shaped (ligulate) in the marginal or ray flowers. Stamens 5, attached on the corolla and mostly united by their anthers into a tube. Pistil with a 1-celled, 1-

seeded, inferior ovary ripening into an achene which is often crowned by the pappus. Seeds without endosperm. — A large family of terrestrial members with only a few aquatics.

KEY TO GENERA

1. Leaves alternate .... *Cotula*
1. Leaves opposite.
    2. Leaves compound or dissected .... *Bidens*
    2. Leaves simple.
        3. Margin entire .... *Jaumea*
        3. Margin serrate or dentate .... *Bidens*

BIDENS

Annual or perennial herbs with opposite leaves. Heads many-flowered, with all flowers tubular (or the marginal ones ligulate and neutral), yellow. Involucre with 2 series of bracts, the outer large and foliaceous. Receptacle nearly flat, with deciduous chaff. Achenes slender, flattened, 4-sided or, rarely, terete, usually crowned with 2 to 6 awns armed with recurved barbs or rarely naked. — A large genus with many weedy species on dry land or in low, wet places; only a few occur in water.

KEY TO SPECIES OF BIDENS

1. Stems erect; leaves lanceolate, toothed; heads numerous, cylindric; ray flowers small or wanting; achenes flat, with 2 awns .... *B. bidentoides*
1. Stems lax; leaves mostly submersed, dissected into capillary segments; heads solitary, flattened; ray flowers showy; achenes terete, with 4 to 6 awns .. *B. beckii*

Bidens beckii Torr. Water marigold. Figs. 149, A–B; 102a, C. Map 372. Submersed in ponds, lakes, and slow streams. Northern states.

Bidens bidentoides (Nutt.) Britt. Beggar-ticks. Fig. 149, C–D. Map 373. On tidal mud flats and shores. Along the North Atlantic Coast.

MAP 372. *Bidens beckii.*

MAP 373. *Bidens bidentoides.*

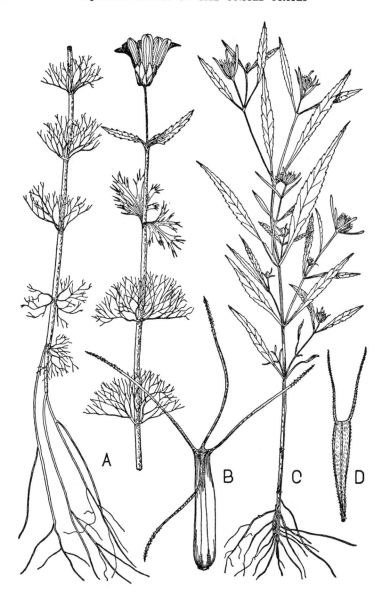

Fig. 149. *Bidens beckii* (A, B) and *B. bidentoides* (C, D).
(A) Plant in blossom; x 1/2. (B) Achene; x 3.
(C) Plant; x 1/2. (D) Achene; x 3.

THE PLANTS ARRANGED BY FAMILIES 337

FIG. 150. *Cotula coronopifolia* (A, B) and *Jaumea carnosa* (C).
(A) Plant; x 1/2. (B) Leaf; x 1.
(C) Plant; x 1/2.

COTULA

Perennial or annual strong-scented herbs with erect, decumbent, or creeping, somewhat succulent stems. Leaves alternate, pinnately dissected, lobed, or sometimes the upper or all of them nearly entire, clasping at the base. Heads terminal on slender peduncles, up to 1 cm. in diameter, depressed-hemispheric, many-flowered. Involucre of green, scarious-margined bracts; receptacle flat or convex, naked. The marginal flowers reduced, pistillate, and fertile; those on the disk, tubular, perfect or staminate, bright yellow. Achenes raised on stalks which remain on the receptacle, with corky wings; pappus wanting.

Cotula coronopifolia L. Brass buttons. Fig. 150, A–B. Map 374.

Locally forming dense mats in depressions in salt marshes and on muddy shores of tidal streams. Believed to have been introduced from South Africa. Naturalized along the coast from Washington to California; reported as adventive on the coast of New England.

JAUMEA

Perennials with weak or ascending, succulent branches from prostrate or creeping stems. Leaves opposite, entire, connate at base, linear or nearly so, fleshy, about 2 to 4 cm. long. Heads solitary and terminal on the branches. Involucre narrowly campanulate, the bracts broad, the outer short and fleshy. Receptacle conical, naked. Ray flowers 9 to 12, pistillate, narrow, yellow; disk flowers several, perfect. Achenes linear, 10-nerved; pappus wanting.

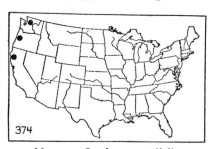
Map 374. *Cotula coronopifolia*.

Map 375. *Jaumea carnosa*.

Jaumea carnosa (Less.) Gray. Fig. 150, C. Map 375.

Local in salt marshes and along tidal streams along the seacoast. Washington to California.

REFERENCE

Sherff, E. E. The genus *Bidens*. Field Mus. Nat. Hist. Bot. Ser. 16:1–709. 1937.

ISOËTACEAE: Quillwort Family

Small, aquatic or semiaquatic, perennial herbs with short, cormlike stem crowned with a tuft of quill-like leaves. The roots arise from the 2- to 4-lobed base of the "corm." Spores of two kinds, borne in axillary sporangia. The single genus, *Isoëtes,* with about 75 species, is restricted chiefly to the temperate regions.

ISOËTES: Quillwort

Leaves 5 to 200, crowded in a fascicle-like, close spiral, bright green, olive-green, or reddish. In mature plants the leaves are from 2 cm. to 1 meter long, slender and tapering, angular or nearly terete, rigid and erect or recurved, or flaccid and floating on the surface of the water. Emersed leaves especially may have bands of stomates; in cross section 4 air tubes, a central bast bundle, and, in some species, 4 or more peripheral bast bundles may be visible.

Sporangia 4 to 7 mm. long, of two kinds, solitary, sunken in the flattened or concave ventral surface of the base of the leaf. The microsporangia, borne mostly on the inner leaves, produce microspores from 20 to 45 microns in diameter. The megasporangia, borne mostly on the outer leaves or sometimes on separate plants, produce megaspores from 250 to 900 microns in diameter; when mature these are variously sculptured over their surface with spines, tubercles, ridges, or reticulations. These spore markings are used in the determination of species. The thin edges of the leaf form a kind of flap, the velum, that partially or sometimes more or less completely covers the sporangium.

Note. Some few species of Isoëtes may grow submersed to a depth of 4 or 5 meters, even forming a dominant part of the vegetation on stony or gravelly shoals in some of the larger, clear-water lakes; others are strictly terrestrial or grow in situations where inundation is very infrequent. Most of the American species grow in shallow waters where they are inundated most of the time but emersed or stranded when the water recedes. Here they thrive in the zone between the high- and low-water lines. In many of the man-made lakes or reservoirs of the northeastern states the areas included in this zone, made by the alternate impounding and withdrawal of the water, are very extensive. The "amphibious" species of *Isoëtes* thrive in such habitats.

In some of the reservoirs of the metropolitan New York water supply, thousands of acres are covered with grassy meadows of *Isoëtes*. They grow submersed, and when the water is withdrawn the exposed areas are covered with a green "turf" that prevents erosion of the shore and silting or roiling of the water. Sometimes plants exposed for a long time die back and later in the season, or after rains, produce a new growth of leaves quite unlike those of the submersed state. Cattle and deer have been observed to feed on the leaves, and muskrats and waterfowl sometimes eat the fleshy "corms" of *Isoëtes*.

KEY TO SPECIES OF ISOËTES

1. Sculpture on the surface of megaspores composed of spines or tubercles.
   2. Megaspores spiny; stomata present or lacking .... *I. echinospora*
   2. Megaspores tuberculate.
      3. Velum complete or nearly so.
         4. "Corm" usually 3-lobed .... *I. orcuttii*
         4. "Corm" 2-lobed.
            5. Megaspores dark when wet .... *I. melanospora*
            5. Megaspores white or cream-colored .... *I. flaccida*
      3. Velum narrow, covering not more than ⅓ of the sporangium.
         6. Tubercles on megaspores simple.
            7. Megaspores less than 480 microns in diameter .... *I. melanopoda*
            7. Megaspores more than 480 microns in diameter .... *I. butleri*
         6. Tubercles on megaspores confluent.
            8. Plants amphibious; leaves with peripheral strands .... *I. howellii*
            8. Plants submersed; leaves lacking peripheral strands ...... *I. bolanderi*
1. Sculpture on the surface of megaspores composed of irregular crests or reticulations; corms usually 2-lobed.
   9. Megaspores reticulated, at least on the basal face.
      10. Megaspores 600 to 800 microns in diameter .... *I. macrospora*
      10. Megaspores 400 to 600 microns in diameter.
         11. Plants amphibious; stomata common, at least near the leaf tips ....
            .... *I. engelmanni*
         11. Plants submersed; stomata rare or absent .... *I. tuckermani*
   9. Megaspores irregularly crested.
      12. Plants submersed; leaves lacking stomata .... *I. occidentalis*
      12. Plants amphibious; leaves with numerous stomata.
         13. Megaspores covered with crowded, irregular ridges.
            14. Leaves 25 to 200, mostly with peripheral strands .... *I. eatoni*
            14. Leaves 8 to 25, without peripheral strands .... *I. saccharata*
         13. Megaspores covered with more distant, jagged crests .... *I. riparia*

Isoëtes bolanderi Engelm. Map 376.

In alpine lakes and ponds. From the Rocky Mountains westward to the Cascades and Sierra Nevadas. Much-reduced plants with short

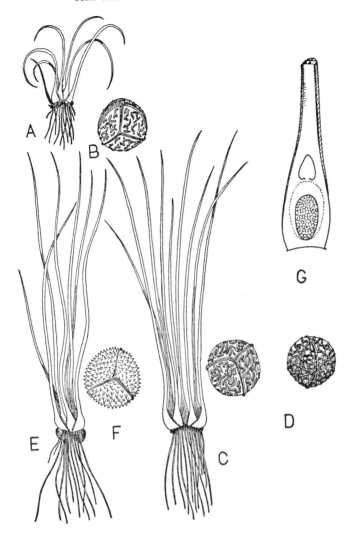

Fig. 151. *Isoëtes tuckermani* (A, B), *I. engelmanni* (C, D), and *I. echinospora* (E, F).
(A) Plant; x 1. (B) Megaspore; x 30.
(C) Plant; x 1/4. (D) Megaspore; x 30.
(E) Plant; x 1/3. (F) Megaspore; x 30.
(G) General diagram of basal part of leaf showing velum; x 2.

leaves occurring in the high mountain lakes along the Nevada-California boundary and in Arizona have been described as *I. bolanderi* var. *pygmaea* Clute.

Isoëtes butleri Engelm. Map 377.

Local on alkaline flats and in springy places on limestone areas. From Tennessee to Kansas and Oklahoma.

Isoëtes eatoni Dodge. Map 378.

In shallow ponds and lake margins. In the northeastern states.

Isoëtes echinospora Dur. Fig. 151, E–F. Map 379.

Usually submersed on sandy or gravelly bottom, sometimes in deep water, in lakes, ponds, and sluggish streams. Throughout the northeastern states and also from the northern Rocky Mountains to the Pacific Northwest. This is the most common and widespread species occurring in the United States. Most of the American representatives of this species are usually assigned to *I. echinospora* var. *braunii* Engelm. Several variations of this species have been given names. A robust form with many long leaves, frequently floating on the surface of streams and outlets of ponds in Vermont, Connecticut, and New York, has been designated *I. gravesii* Eaton (*I. echinospora robusta* Pfeiffer). The specimens from Spanaway Lake, Washington, described as *I. echinospora* var. *flettii* Eaton or *I. flettii* Pfeiffer on the basis of possessing blunt tubercles or short ridges instead of sharp spines, appear to belong here.

MAP 376. *Isoëtes bolanderi.*

MAP 377. *Isoëtes butleri.*

MAP 378. *Isoëtes eatoni.*

MAP 379. *Isoëtes echinospora.*

Map 380. *Isoëtes engelmanni*.

Map 381. *Isoëtes flaccida*.

Map 382. *Isoëtes howellii*.

Map 383. *Isoëtes macrospora*.

Isoëtes engelmanni A. Br. Fig. 151, C–D. Map 380.
In intermittent ponds and shallow borders of lakes and sluggish streams where they become exposed at low water. Common in the eastern United States. Specimens from Georgia and North Carolina, with a velum covering nearly ⅔ of the sporangium instead of only ⅓, have been separated as *I. engelmanni* var. *caroliniana* Eaton.

Isoëtes flaccida Schuttlw. Map 381.
In marshes or shallow ponds and streams, often emersed. Mostly in the pine barrens, Georgia and Florida.

Isoëtes howellii Engelm. Map 382.
In shallow lakes and ponds. From western Montana to the Pacific Coast.

Isoëtes macrospora Dur. Map 383.
On sandy bottom in shallow water of lakes and streams. From New England to the Great Lakes states.

Isoëtes melanopoda Gay. and Dur. Map 384.
In shallow ponds, sloughs, and wet meadows subject to desiccation in late summer. Mostly in the central part of the Mississippi Basin.

Isoëtes melanospora Engelm. Map 385.
In shallow pools and depressions on Stone Mountain, Georgia.

Isoëtes occidentalis Henders. Map 386.

Submersed in shallow or deep water on sandy bottom of mountain lakes. From Colorado to the Pacific Coast. The specimens from Washington have been described as a separate species, *I. piperi* Eaton, on the basis of their tendency to produce slightly larger megaspores and slenderer and somewhat longer leaves. Such slight and inconstant differences hardly justify the separation of the plants of one state from those of contiguous states as a separate species.

Isoëtes orcuttii Eaton. Map 387.

In pools and wet meadows in low depressions submersed during

Map 384. *Isoëtes melanopoda.*

Map 385. *Isoëtes melanospora.*

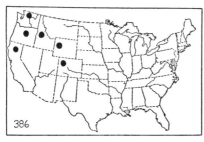

Map 386. *Isoëtes occidentalis.*

Map 387. *Isoëtes orcuttii.*

Map 388. *Isoëtes riparia.*

Map 389. *Isoëtes saccharata.*

much of the growing stages but subject to desiccation. Southern California. This species is similar to, but smaller than, the apparently closely related terrestrial *I. nuttallii* A. Br. of the Pacific Coast states.

Isoëtes riparia Engelm. Map 388.

Mainly on muddy shores and flats of tidal streams and estuaries, occasionally inland. North Atlantic states. A more robust form with larger and more numerous leaves has been described as *I. riparia* var. *canadensis* Engelm. (*I. dodgei* Eaton).

Isoëtes saccharata Engelm. Map 389.

On shores and mud flats of tidal streams and estuaries from New Jersey to Virginia.

Isoëtes tuckermani A. Br. Fig. 151, A–B. Map 390.

Submersed in shallow ponds, lakes, and streams, mostly on gravelly bottom. New England and New York.

Map 390. *Isoëtes tuckermani.*

REFERENCES

Pfeiffer, N. E. Monograph of the Isoëtaceae. Ann. Mo. Bot. Gard. 9:79–232. 1922.
——. A new *Isoëtes* from Virginia. Claytonia 3:29–30. 1937.
——. A new variety of *Isoëtes virginica*. Torrey Bot. Club Bull. 66:411–413. 1939.
Iversen, Johannes. Über die Species-umgrenzung und Variation der *Isoëtes echinospora* Dur. Bot. Tidskrift 40:126–131. 1928.
Erickson, L. C. A study of *Isoëtes* in San Diego County, California. Madroño 6:7–11. 1941.

EQUISETACEAE: Horsetail Family

Rushlike herbs with hard, jointed, simple or branched stems from creeping rootstocks. Leaves in whorls, small, simple, and fused laterally into a sheath or collar around the node. — The single genus,

Equisetum, contains about 25 species, nearly all of which are terrestrial in the cooler temperate regions.

EQUISETUM: Horsetail

Rootstocks perennial, creeping, with fibrous roots at the nodes, frequently tuber-bearing. Stems annual or perennial, all alike, or the fertile succulent and chlorophyll-less, and the sterile green, usually with whorls of branches; both striated and usually with a central cavity. Sporangia borne in terminal cones. Spores with 4 elastic, hygroscopic bands (elaters).

Equisetum fluviatile L. Pipes. Fig. 152, A–D. Map 391.

Stems annual, from 3 to 20 dm. tall, simple or sparingly branched. Leaf sheath with rigid, sharp, dark-brown teeth.

While several species may be found in wet, marshy lands, this is the only one that actually thrives in water with only the upper part of the plant emersed. In shallow lakes and ponds and on muddy shores and stream banks. Common across the northern United States.

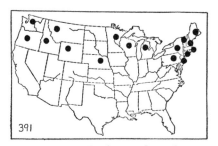

Map 391. *Equisetum fluviatile.*

MARSILEACEAE: Water-Clover Family

Perennial herbs, mostly with slender, branching, creeping rootstocks bearing roots at the nodes. Leaves alternate, in 2 ranks, compound or simple. Sporocarps hard, nutlike or bean-shaped, bearing two kinds of sporangia. — The family contains 3 genera, of which only *Marsilea* is represented by aquatic species in the United States. *Pilularia americana* A. Br., with simple, linear leaves, occurs in wet depressions in several of the southern and western states.

MARSILEA: Water Clover

Leaves with slender petioles, bearing 4 leaflets with entire or slightly undulate margins. Sporocarps several or solitary on a peduncle, the

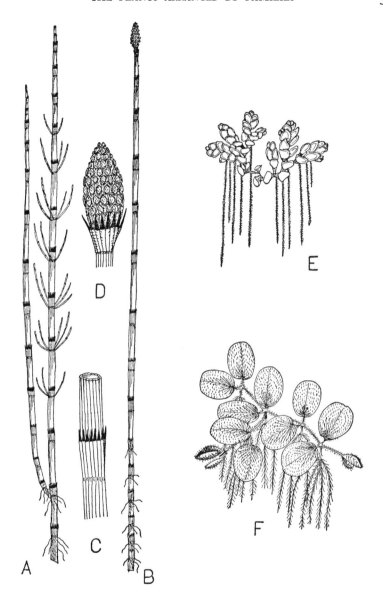

Fig. 152. *Equisetum fluviatile* (A–D), *Azolla caroliniana* (E), and *Salvinia rotundifolia* (F).

(A) Sterile shoots; x 1/3. (B) Fertile shoot; x 1/3. (C) Stem node with sheath of fused leaves terminating in whorl of teeth; x 1. (D) Cone; x 1.
(E) Plant; x 2.
(F) Plant; x 1.

latter axillary or adnate to the lower part of the petiole. Sporocarps with a prominent suture with 2 teeth near the base, dehiscing with age into 2 valves each emitting a gelatinous mass with from about 6 to 14 sori. Each sorus contains numerous microsporangia and a few megasporangia.

NOTE. *Marsilea* plants growing in deep water may produce petioles a meter long with leaflets floating on the water surface. Such plants produce few sporocarps. Plants growing in ponds or streams with fluctuating water levels may produce a very dense mat of creeping stems with clusters of many erect, short-petioled leaves. Such emersed plants may produce large numbers of sporocarps in late summer.

KEY TO SPECIES OF MARSILEA

1. Peduncles with 2 to 6 sporocarps, adnate to the base of the petiole; basal teeth of sporocarp minute.
    2. Leaflets glabrous; mature sporocarp glabrescent .... *M. quadrifolia*
    2. Leaflets silky with white hairs; mature sporocarp tomentose or villous .... *M. macropoda*
1. Peduncles with solitary sporocarp (rarely with 2), not adnate to the petiole; basal teeth of sporocarp prominent.
    3. Peduncle short, not exceeding length of sporocarp; basal teeth not hooked .... *M. vestita*
    3. Peduncle at least as long as the sporocarp; upper basal tooth hooked .... *M. uncinata*

Marsilea macropoda Engelm. Fig. 153, E. Map 392.
Along streams and in wet depressions. Southern Texas.

Marsilea quadrifolia L. Fig. 153, A–B. Map 393.
In shallow ponds, bays of lakes, and sluggish streams. Introduced from Europe and well established in several localities in the northeastern states; also frequently grown in garden pools.

Marsilea uncinata A. Br. Fig. 153, D. Map 394.
In ponds and wet swales. Texas and Louisiana.

Marsilea vestita Hook. and Grev. Fig. 153, C. Map 395.
Locally common in shallow water, on shores of streams and ponds, and in wet depressions. From the Great Plains westward; introduced locally in the East.

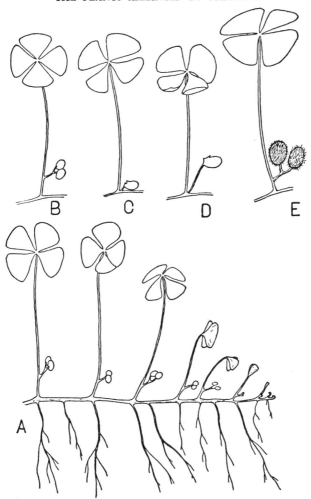

Fig. 153. *Marsilea quadrifolia* (A, B), *M. vestita* (C), *M. uncinata* (D), and *M. macropoda* (E).

(A) Part of plant with sporocarps; x 1/3. (B–E) Part of stem showing attachment of leaf and adjacent stalk with sporocarps; x 1/3.

Map 392. *Marsilea macropoda*.

Map 393. *Marsilea quadrifolia*.

MAP 394. *Marsilea uncinata.*

MAP 395. *Marsilea vestita.*

SALVINIACEAE: WATER-FERN FAMILY

Small, floating plants with an elongate, horizontal, simple or branched axis. Sporocarps thin-walled, soft, 2 or more on a peduncle, containing either megasporangia with solitary megaspores or microsporangia with numerous microspores. — There are 2 genera, *Salvinia*, with 13 species, and *Azolla*, with 5 species, all inhabiting tropical or warm temperate regions.

KEY TO GENERA

1. Leaves several, about 1 to 2 cm. long, with a distinct midrib; roots absent but the lower leaves dissected into filiform, hairy segments .... *Salvinia*
1. Leaves numerous, minute, overlapping, deeply 2-lobed; roots on the lower side of stem .... *Azolla*

AZOLLA: WATER FERN

Plants delicate, mosslike, with pinnately branched axis covered with small, scalelike, 2-lobed leaves. Leaves in 2 ranks, the upper lobe of each leaf aerial, the lower submersed. Sporocarps in clusters of 2 or 4 on the submersed lobe of the first leaf of a lateral branch, unequal in size, those bearing microsporangia large, those bearing the solitary megasporangia, small.

NOTE. Propagation by fragmentation of the plants is rapid. Sometimes a whole pond or large area of water is completely covered with a blanket of floating plants. Young plants are at first a bright or gray-green but with age these may turn pink, red, or dark brown. The lower surfaces of the upper lobes of leaves have large cavities in which colonies of the blue-green alga, *Anabaena*, thrive in profusion.

KEY TO SPECIES OF AZOLLA

1. Megaspores coarsely roughened; masses of microspores with nonseptate processes; plants usually from 1 to 10 cm. long; branches numerous, open ....
.... *A. filiculoides*
1. Megaspores finely roughened; masses of microspores with septate processes; plants mostly less than 3 cm. long; branches fewer, more crowded .. *A. caroliniana*

Azolla caroliniana Willd. Fig. 152, E. Map 396.

Locally common in shallow ponds, canals, ditches, and sluggish streams. From New England to the Pacific Coast.

Azolla filiculoides Lam. Map 397.

In shallow ponds and sluggish streams. California; Long Island, New York; perhaps elsewhere in the South.

Map 396. *Azolla caroliniana.*

Map 397. *Azolla filiculoides.*

SALVINIA: Water Fern

Leaves arranged on the axis in whorls of 3; the 2 lateral floating, oblong to orbicular, 1 to 2 cm. long; the lower submersed, dissected into several filiform, rootlike, hairy segments. The floating leaves are covered with stiff hairs and papillose projections on the upper surface. Sporocarps 4 to 20 in clusters on the segments of submersed leaves, all alike in size, the first 1 or 2 with megasporangia, the others with microsporangia.

Salvinia rotundifolia Willd. (*S. natans* of American manuals). Fig. 152, F. Map 398.

This is a common introduced plant grown in aquaria and pools. It usually does not survive the winters in the northern states. In several localities it has become naturalized, chiefly in shallow, warm pools; possibly native in the South. It is common in lakes and sluggish streams in Central America.

MAP 398. *Salvinia rotundifolia*.

REFERENCES

Baker, J. G. Handbook of the fern allies. 159 pp. London, 1887.
Herzog, Robert. Ein Beitrag zur Systematik der Gattung *Salvinia*. Hedwigia 74:257–284. 1935.
Weatherby, C. A. A further note on *Salvinia*. Amer. Fern Jour. 27:98–102. 1937.
Clausen, R. T. *Azolla filiculoides* on Long Island [N.Y.]. Amer. Fern Jour. 30:103. 1940.

PARKERIACEAE: FLOATING-FERN FAMILY

Plants with short, horizontal rootstocks; roots mostly from the basal part of the stipes; fronds dimorphic, in a rosette-like cluster. Sporangia globose, in 1 or 2 rows along the longitudinal veins near the revolute margin of the frond; annulus complete, partial, or obsolete; indusia formed of the reflexed margins of the frond.—A single genus, *Ceratopteris*, with about 4 species, limited to tropical and subtropical regions. It has sometimes been placed in the Polypodiaceae and it has also been made the basis of the family Ceratopteridaceae.

CERATOPTERIS: FLOATING FERN, HORN FERN

Plants floating or sometimes rooted in the mud. Sterile fronds fleshy, from 1 to 4 dm. long, at first simple, ultimately pinnate or twice-pinnate, with rounded lobes or apex, frequently bearing buds in the notches along the margin; fertile fronds much exceeding the sterile; several times pinnately divided into linear segments.

KEY TO SPECIES OF CERATOPTERIS

1. Stipes bulbous; sterile fronds mostly floating; sporangia containing 32 spores .... *C. pteridoides*
1. Stipes not bulbous; sterile fronds mostly emersed; sporangia containing 16 spores .... *C. deltoides*

Fig. 154. *Ceratopteris pteridoides.*
A small sterile plant showing new plants forming from buds in notches of leaf margins; x 1/3.

Ceratopteris deltoides Benedict. Map 400.

Floating on ponds or sluggish streams, sometimes rooted in mud along their shores. Florida, Louisiana, and possibly elsewhere along the Gulf Coast. In cultivation sometimes confused with the following species.

Map 399. *Ceratopteris pteridoides.*   Map 400. *Ceratopteris deltoides.*

Ceratopteris pteridoides (Hook.) Hieron. Fig. 154. Map 399.

Locally common, floating on ponds and sluggish streams. In southern Florida; also reported from the Gulf Coast of Mississippi and

Louisiana. In the northern states it is grown in aquaria and conservatory pools. It thrives in the open during hot weather but does not survive the northern winter.

REFERENCES

Benedict, R. C. Ceratopteridaceae. In, North Amer. Flora 16 (1):29-30. 1909.
Brown, C. A., and D. S. Correll. Ferns and fern allies of Louisiana. 186 pp. Baton Rouge, 1942.

GLOSSARY

Acaulescent. Stem apparently wanting or very short.
Achene. Small, dry, hard, one-seeded, indehiscent fruit.
Acuminate. Gradually tapering to a long point.
Acute. Sharp-pointed.
Adnate. Characterized by congenital union of two different organs.
Adventitious buds. Buds appearing in irregular places or order, as those appearing about a wound.
Aggregate fruit. A fruit formed by the coherence of several pistils that were distinct in the flower.
Albuminous seed. A seed containing an endosperm or albumen, the material surrounding the embryo.
Alternate. Having one leaf or bud at a node; placed singly at different heights on the stem.
Angiospermae. Plants with seeds borne in an ovary.
Annual. Of one season's duration from seed to maturity and death.
Annulus. A ring or band, as in sporangia of ferns.
Anther. The pollen-bearing part of the stamen.
Apetalous. Having no petals.
Apiculate. Ending in a short, pointed tip.
Appressed. Lying close and flat against.
Areolate. Having the surface marked off into small areas like the meshes of a net.
Aril. An appendage or covering of a seed growing from its hilum or base.
Articulate. With joint or node.
Ascending. Rising obliquely upwards.
Attenuate. Becoming very narrow.
Auriculate. With ear-shaped appendages.
Awl-shaped. Narrow and sharp-pointed; gradually tapering from the base to a slender or stiff point.
Awn. A bristle-like part or appendage.
Axil. The upper angle formed where a leaf joins a stem.
Axile. Situated on the axis.
Axillary. Situated in an axil.
Axis. The main or central line of development of a plant or organ; the stem.

Barbed. Furnished with rigid points or short bristles, usually reflexed like the barb of a fishhook.
Beak. A long, prominent, and substantial point. Usually applied to prolongations of fruits and pistils.
Bearded. Bearing a long awn or furnished with long or stiff bristles.
Berry. A fleshy fruit, soft throughout.
Bi-. A Latin prefix signifying *two, twice,* or *doubly.*
Biennial. Of two seasons' duration from seed to maturity and death.
Blade. The expanded part of a leaf.
Bract. A much-reduced leaf, particularly one of the small or scalelike leaves of a flower cluster or one of those associated with the flowers.
Bracteate. Having bracts.
Bractlet. A secondary bract, as one borne on the pedicel of a flower.
Bulb. An underground leaf bud with fleshy scales and a short axis.
Bulblet. A small bulb borne on the stem or inflorescence.
Bulbous. Having the character of a bulb.
Callus. The hard swelling at the base or insertion of the lemma or palea in grasses.
Calyx. The outer whorl of floral envelopes; the outer perianth whorl.
Campanulate. Bell-shaped; cup-shaped with a broad base.
Canescent. Hoary with gray pubescence.
Capillary. Hairlike; very slender.
Capitate. Shaped like a head; aggregated into a dense or compact cluster or head.
Capsule. Dry, dehiscent fruit of a compound pistil.
Carinate. Having a keel or projecting longitudinal medial line on the lower or outer surface.
Carpel. One of the foliar units of a compound pistil. A simple pistil has one carpel.
Caryopsis. A grain.
Caudate. With a tail-like appendage.
Cauline. Pertaining or belonging to the stem.
Cell. A structure containing a cavity, as the cells of an ovary or anther.
Cespitose. Forming tufts or clusters.
Chaff. A small, thin scale or bract, becoming dry or membranous; in particular the bracts in the flower heads of composites.
Chlorophyll. The green coloring matter within the cells of plants.
Ciliate. Fringed with hairs on the margin.
Cinereous. Ash-color; light gray.
Circumscissile. Dehiscing by a regular, transverse, circular line of division, the top usually separating as a lid.
Clasping. Of leaves, partly or completely surrounding a stem.
Claw. The long, narrow, petiole-like base of a petal or sepal in some flowers.
Cleft. Lobed, with the incisions extending halfway or more to the midrib of a leaf.
Cleistogamous. Of flowers, producing seed without opening the buds for fertilization.
Comose. Furnished with or like a tuft of hairs.
Compound leaves. Leaves in which the blade consists of two or more separate parts (leaflets).
Compound pistil. A pistil composed of two or more fused carpels.

Connate. Characterized by congenital union of like structures.
Connective. The part of a stamen between its two anther cells.
Connivent. Coming together or converging but not organically connected.
Cordate. Heart-shaped, with the point away from the base.
Coriaceous. Of the texture of leather.
Corm. A solid, bulblike part usually underground; the enlarged, fleshy, solid base of a stem.
Corolla. The inner whorl of floral envelopes; the inner perianth whorl.
Corymb. A flat or convex flower cluster with the outer flower opening first.
Cotyledon. A seed leaf; the primary leaf or leaves in the embryo.
Creeping. Running along the ground and rooting its entire length, as of stems.
Crenate. With rounded teeth; scalloped.
Crested. With an elevated and irregular or toothed ridge.
Cucullate. With or like a cowl or hood.
Culm. The stem of grasses or sedges.
Cuneate. Wedge-shaped; triangular with the acute angle downward.
Cuspidate. Tipped with a sharp and rigid point.
Cyme. A broad, more or less flat-topped, flower cluster, with central flowers opening first.
Deciduous. Falling off in autumn, as leaves; falling early.
Decompound. More than once compound or divided.
Decumbent. Reclining at base but the summit ascending.
Decurrent. Extending down the stem below the insertion.
Deflexed. Bent or turned abruptly downward.
Dehiscent. Opening regularly by valves or slits, as a capsule.
Deltoid. Triangular; delta-like.
Dentate. Toothed, with teeth pointing outward.
Denticulate. Furnished with minute teeth.
Depressed. Somewhat flattened from the end or top.
Di-. A Greek prefix signifying *two* or *twice*.
Diadelphous. Of stamens, combined in two sets or groups.
Dichotomous. Forking regularly by pairs.
Dicotyledon. A plant of the subdivision of Angiospermae bearing two cotyledons or seed leaves.
Diffuse. Loosely branching or spreading; of open growth.
Digitate. Handlike; compound with the members arising from a common point.
Dimorphous. Having two forms.
Dioecious. Unisexual with staminate and pistillate flowers on different plants.
Disk. A more or less fleshy or elevated development of the receptacle about the pistil; a receptacle in the heads of composites.
Disk flowers. The tubular flowers in the center of the heads of composites, as distinguished from the marginal or ray flowers.
Dissected. Divided into many slender segments.
Divaricate. Widely divergent or spreading.
Divided. Separated to the base.
Dorsal. Back; relating to the back or outer surface of a part of an organ.
Drupe. Fleshy, indehiscent fruit with endocarp stony and exocarp soft; stone fruit.

Drupelet. A small drupe.
Elliptical. Having the shape of an ellipse; oval or oblong with the ends rounded.
Emarginate. With a shallow notch at the apex.
Emersed. Growing or extending above the surface of the water.
Endocarp. The inner layer of a pericarp.
Endosperm. Nutritive tissue or material about the embryo in a seed.
Entire. Having an even margin; not toothed, notched, or divided.
Ephemeral. Short-lived; persisting for one day only.
Epigynous. Borne on the ovary. Used of floral parts when the ovary is inferior and the flower not perigynous.
Erose. Having an uneven or notched margin as if torn.
Exalbuminous seed. Seed not having an albumen or endosperm.
Exocarp. The outer layer of a pericarp.
Exserted. Projecting beyond an envelope.
Extrorse. Turned or facing outward or away from something.
Falcate. Scythe-shaped or sickle-shaped.
Fascicle. A condensed or close cluster.
Fastigiate. Having stems or branches which are nearly erect or close together.
Fertile. Of stamens, bearing pollen; of flowers, bearing seeds.
Filiform. Threadlike; long and very slender.
Fimbriate. Fringed.
Flexuous. Having a more or less zigzag or wavy form.
Floret. An individual flower of grasses and composites; also, any other very small flower that forms a part of a dense inflorescence.
Follicle. A fruit consisting of a single carpel splitting along the inner or upper suture only.
Free-central placentation. An arrangement of the placenta in which it projects from the bottom of the ovary but is otherwise free.
Frond. The leaf of ferns; the plant body of Lemnaceae.
Fruit. The ripened ovary or ovaries with the adnate parts; the seed-bearing organ.
Fugacious. Falling or fading soon; of short duration.
Fusiform. Spindle-shaped; swollen in the middle and tapering toward each end.
Gamopetalous. Having the petals united; having a corolla of one piece.
Gibbous. Swollen on one side.
Glabrate. Somewhat glabrous, or becoming glabrous with maturity.
Glabrous. Without hairs.
Gland. A secreting part or appendage. The term is often used for small swellings or projections on various organs.
Glandular. Furnished with glands or glandlike.
Glaucous. Covered or whitened with a bloom.
Globose. Spherical in form or nearly so.
Globular. Nearly globose.
Glume. One of the two basal bracts in a grass spikelet, without a flower in its axil.
Glutinous. Sticky; gluey.
Grain. A fruit like an achene but the seed coat and thin pericarp are fused throughout into one body; particularly the fruit of grasses.
Granular, granulose. Composed of, or appearing as if covered by, minute grains.

Habit. The general aspect of a plant or its mode of growth.
Habitat. The home of a plant; the situation in which a plant grows wild.
Halberd-shaped, hastate. Shaped like an arrowhead but with the basal lobes spreading.
Head. A short, compact flower cluster of more or less sessile flowers.
Herb. A plant naturally dying to the ground; one without persistent stem above ground; one without a definite woody structure.
Herbaceous. Not woody; of the texture of an herb.
Hilum. In the seed, the scar or mark indicating the point of attachment.
Hirsute. With rather coarse or rough hairs.
Hispid. Provided with stiff or bristly hairs.
Hyaline. Transparent or translucent.
Hypogynous. Borne on the axis or under the ovary. Said of stamens or petals and of sepals.
Imbricated. Overlapping like the shingles on a roof.
Imperfect flower. A flower lacking either stamens or carpels.
Included. Not protruded, as stamens not projecting from the corolla; not exserted.
Indehiscent. Not regularly opening, as a seed pod or anther.
Indurated. Hardened.
Indusium. The covering of the sorus or fruit dot in ferns.
Inferior ovary. An ovary in which the sepals appear attached to the top.
Inflexed. Bent inward.
Inflorescence. A flower cluster; a mode of flower-bearing.
Internode. The part of the stem between two nodes.
Involucel. An involucre about a part of a flower cluster; a secondary involucre.
Involucre. A whorl of small leaves or bracts standing close below a flower or flower cluster.
Involute. Rolled or coiled inward.
Irregular flower. A flower having parts of a whorl or series not all alike.
Keeled. Ridged like the bottom of a boat.
Lacerate. With the margin appearing as if torn.
Laciniate. With narrow, irregular lobes.
Lanceolate. Several times longer than wide, broadest near the base and narrowed to the apex.
Lateral. On or at the side.
Leaflet. One part of a compound leaf.
Lemma. One of the bracts of the grass spikelet bearing a flower in its axil.
Lenticular. Lens-shaped; with two convex surfaces.
Ligulate. Strap-shaped. Used particularly of the ray flowers of composites.
Ligule. A projection or outgrowth from the top of the sheath in grasses and similar plants.
Limb. The expanded part of a petal or corolla.
Linear. Long and narrow with parallel margins.
Lobed. Divided into segments about to the middle.
Loculicidal dehiscence. Splitting through the back of each cell or carpel of a capsule.
Lyrate. Pinnatifid but with an enlarged terminal lobe and smaller lateral lobes.
Maculate. Spotted.

Marginal flowers. The outer flowers, particularly in the head of composites, either ligulate or tubular.

Megaspore. The larger of two kinds of spores in heterosporous plants.

Membranaceous. Thin, rather soft, and somewhat translucent.

Mericarp. One of the two carpels of the fruit of a member of the Parsley family.

-Merous. Referring to the number of parts; as flowers 5-merous, in which the parts of each kind or series are five or in fives.

Micropyle. The point on the seed which represents the closed orifice of the ovule.

Microspore. The smaller of two kinds of spores in heterosporous plants.

Monadelphous. Having stamens united into one group by their filaments.

Moniliform. Like a string of beads.

Monocotyledon. A plant of the subdivision of Angiospermae bearing one cotyledon or seed leaf.

Monoecious. Having staminate and pistillate flowers on the same plant.

Mucronate. Furnished with an abrupt, minute point.

Multiple fruit. Fruit formed by the coherence of pistils and associated parts of the several flowers of an inflorescence.

Naked. Of a flower, without calyx or corolla; of a receptacle of composites, without chaff.

Nerve. A simple vein or slender rib.

Netted-venation. An arrangement of leaf veins in which the principal veins form a network.

Neutral flower. A flower without stamens or carpels.

Node. A joint or place where leaves are attached to a stem or branch.

Nut. A hard, indehiscent, one-celled, and one-seeded fruit, usually resulting from a compound ovary.

Nutlet. A small nut.

Ob-. A Latin syllable usually indicating inversion; as *obovate,* inverted ovate.

Oblanceolate. Lanceolate but with the narrow end towards the stem.

Oblique. Of leaves, having unequal sides.

Oblong. Longer than broad and with the sides nearly parallel most of their length.

Obtuse. Blunt or rounded at the end.

Ocrea. A sheath formed by the fused stipules, as in some Polygonaceae.

Opposite. Having two leaves or buds at the node.

Orbicular. Circular.

Ovary. The part of the pistil bearing the ovules.

Ovate. With an outline like that of a hen's egg, with the broad end toward the base.

Ovoid. A solid that is oval in flat outline.

Ovule. An undeveloped seed.

Palea. In the grass flower, the upper of the two inclosing bracts, the lower one being the lemma or flowering glume.

Palmate. Radiating fanlike from approximately one point.

Palmately compound leaves. Leaflets radiating from one point.

Panicle. An elongated, irregularly branched, raceme-like inflorescence.

Papillose. Covered by or bearing many minute, nipple-shaped projections.

Pappus. The modified calyx limb in composites, forming a crown of bristles, scales, etc., at the summit of the achene.

GLOSSARY

Parallel venation. An arrangement of leaf veins in which the principal veins run parallel or nearly so.
Parietal. On the inner wall or surface of a capsule.
Parted. Cleft or cut not quite to the base.
Pectinate. Comblike; pinnatifid with very narrow close divisions or parts.
Pedicel. The stem of an individual flower in a cluster.
Peduncle. The stem of a solitary flower or of a flower cluster.
Pellucid. Clear; nearly transparent.
Peltate. Attached to its stalk inside the margin; shield-shaped.
Perennial. Of three or more seasons' duration.
Perfect flower. A flower having both stamens and carpels.
Perfoliate. Of leaves, with the appearance of having the stem pass through the leaf.
Perianth. The floral envelope considered together. Used mostly for flowers having no clear distinction between calyx and corolla, as in lilies.
Pericarp. The matured ovary.
Perigynium. The sac, often inflated, which covers the ovary or fruit of *Carex*.
Perigynous. Borne around the ovary and not beneath it, as when calyx, corolla, and stamens are borne on the edge of a cup-shaped receptacle.
Persistent. Remaining attached; of leaves, not all falling at the same time.
Petal. One of the divisions of the corolla.
Petiole. The stalk of a leaf.
Phyllodium. A petiole having the form and function of a leaf.
Pilose. Shaggy; with soft hairs.
Pinna. A primary division or leaflet of a compound leaf.
Pinnate. Feather-formed.
Pinnately compound. With the leaflets arranged on each side of a common axis of a compound leaf.
Pinnule. A secondary pinna or leaflet in a pinnately decompound leaf.
Pistil. The ovule-bearing and seed-bearing organ with style and stigma.
Pistillate. Having pistils and no stamens; female.
Placenta. The part or place in the ovary where ovules are attached.
Plicate. Folded into plaits, usually lengthwise, like a closed fan.
Plumose. Plumy; feather-like; with fine hairs, as in the pappus bristles of some composites.
Plumule. The bud or growing point of an embryo plant.
Pod. A general term to designate a dry, dehiscent fruit.
Pollen. The spores or grains borne by the anther which contain the male element.
Poly-. A Greek syllable meaning *many*.
Polygamous. Bearing perfect and imperfect flowers on the same plant.
Polypetalous. Having separate petals.
Procumbent. Trailing or lying flat upon the ground but not rooting.
Prostrate. Lying flat upon the ground.
Pteridophyte. A fern or related plant; a spore-bearing vascular plant.
Puberulent. Minutely pubescent.
Pubescent. Covered with soft, short hairs.
Punctate. With translucent or colored dots, depressions, or pits.
Raceme. A simple flower cluster of pediceled flowers on a common elongated axis.

Radical. Of leaves, appearing to come from the root or from the base of the stem near the ground.
Radicle. The part of the embryo below the cotyledons.
Ray flower. One of the modified flowers of the outer part of the heads of some composites with a straplike extension of the corolla.
Receptacle. The more or less enlarged or elongated end of the peduncle or flower axis.
Recurved. Curved outward or backward.
Reflexed. Bent outward or backward.
Regular flower. A flower having all the parts of a whorl or series alike.
Reniform. Kidney-shaped.
Repand. Wavy-margined.
Reticulate. Having a network of veins.
Retrorse. Bent or curved over, back, or downward.
Revolute. Having margins rolled backward or under.
Rhachilla. A diminutive or secondary axis; in particular, in the grasses and sedges, the axis of the spikelet.
Rhachis. The axis of a spike or compound leaf.
Rhizome. An underground stem; a rootstock.
Rib. A primary vein or nerve in a leaf or similar organ; any prominent elevated line along a body.
Rootstock. An underground stem; a rhizome.
Rosette. A very short stem or axis bearing a dense cluster of leaves.
Rotate. Wheel-shaped.
Rudimentary. Imperfectly developed or in an early stage of development.
Rugose. Wrinkled; generally because of the depressions of the veins in the upper surface of the leaf.
Runner. A slender, trailing stem taking root at the nodes.
Sagittate. Shaped like an arrowhead; triangular with the basal lobes pointing downward.
Salverform. With a slender tube and an abruptly spreading border.
Samara. An indehiscent, winged fruit.
Scabrous. Rough to the touch.
Scales. Dry, appressed, and modified or reduced leaves or bracts.
Scape. A leafless peduncle arising from the ground.
Scapose. Resembling a scape.
Scarious. Thin, dry, and membranaceous.
Scurfy. Covered with small, branlike scales.
Sepal. One of the divisions of a calyx.
Septicidal dehiscence. Splitting through the partitions of a capsule.
Serrate. Having sharp teeth pointing forward.
Serrulate. Finely serrate.
Sessile. Without a stalk.
Setose. Full of bristles.
Sheath. A long or more or less tubular structure surrounding an organ or part.
Simple leaves. Leaves in which the blade is all in one piece.
Sinuate. Wavy-margined.

Sinus. The depression or bay between adjacent lobes.
Sorus (pl. *sori*). The fruit dot or cluster of ferns.
Spadix. A thick or fleshy spike surrounded or subtended by a spathe.
Spathe. The bract or leaf surrounding or subtending a flower cluster or a spadix.
Spatulate. Gradually narrowed downward from a rounded summit.
Spermatophyte. A seed-bearing plant.
Spike. A flower cluster like a raceme but with sessile or nearly sessile flowers.
Spikelet. A secondary spike; a unit of the inflorescence of the grasses.
Spine. A sharp, rather slender, rigid outgrowth.
Spinulose. Provided with small spines.
Sporangium. A spore case.
Spore. A simple reproductive body usually consisting of a single detached cell and containing no embryo, particularly in the ferns and lower plants.
Sporocarp. The fruit body of certain Cryptogams containing sporangia or spores.
Spur. A tubular or hornlike projection from the calyx or corolla; it usually secretes nectar.
Stamen. The pollen-bearing or male organ of a flower.
Staminate. Having stamens and no pistils; male.
Staminodium. A sterile stamen or stamen-like organ without anthers.
Stellate. Star-shaped.
Sterile. Of plants, flowers, or stamens, infertile, barren.
Stigma. The part of the pistil which receives the pollen.
Stipe. The stalk of a pistil; the petiole of a fern leaf.
Stipel. The stipule of a leaflet.
Stipitate. Having a stalk or stipe.
Stipule. A basal appendage of a petiole, usually one of two.
Stolon. A shoot that bends to the ground, taking root at its tip and thus giving rise to a new plant.
Stoloniferous. Bearing runners or shoots that take root
Striate. Marked with fine, longitudinal lines or ridges.
Strict. Very straight and upright.
Strigose. With appressed, sharp, straight, and stiff hairs.
Style. The part of the pistil connecting the ovary and stigma, usually more or less elongated.
Stylopodium. A disklike expansion of the base of the style, particularly in the Parsley family.
Sub-. A Latin prefix usually signifying *somewhat* or *slightly*.
Submersed. Growing under water.
Subtended. Standing below or close to, as a bract below a flower.
Subulate. Awl-shaped; broad at base and narrow and tapering from the base to a sharp, rigid point, the sides generally concave.
Succulent. Juicy; fleshy; soft and thickened in texture.
Superior ovary. An ovary that is free from the calyx or perianth.
Supra-axillary. Borne on the stem some distance above the leaf axil.
Taproot. A root with a stout, tapering body, usually vertical.
Tendril. A thread-shaped organ, as a modified leaf or stem part, by which a plant clings to a support.

Terete. Circular in cross section.
Terminal. At the end of a stem or branch.
Ternate. In threes.
Throat. The opening of a gamopetalous corolla or gamopetalous calyx.
Tomentose. Densely hairy with matted wool.
Translucent. Partially transparent.
Tri-. Three or three times.
Trifoliate. Of three leaflets.
Truncate. Ending abruptly as if cut off transversely.
Tuber. A thickened part; usually an enlarged end of a subterranean stem or a rootstock.
Tubercle. A small tuber; a rounded, protruding body.
Tubular. Hollow and of an elongated or pipelike form.
Turbinate. Shaped like a top; inversely conical.
Turion. A hardened, abbreviated axis or winter bud as in *Potamogeton*.
Umbel. An umbrella-like flower cluster.
Umbellet. A secondary umbel.
Unarmed. Destitute of spines, prickles, or thorns.
Uncinate. Hooked at the tip.
Undulate. With a wavy margin or surface.
Unisexual. Of one sex; staminate or pistillate.
Utricle. A small, bladder-like, one-seeded fruit.
Valve. One of the units or pieces into which a capsule splits; a separable part of a pod.
Veins. The branches of the fibrovascular bundles forming the framework in leaves.
Velum. A membranous flap or covering from the leaf margin over the sporangia, in *Isoëtes*.
Ventral. Front; relating to the inner face of an organ; opposite of dorsal.
Verticil. A small whorl.
Villous. Having long, soft hairs; shaggy but not matted.
Viscid. Sticky.
Whorled. Having leaves or buds arranged in a group of three or more at a node.
Wing. A thin, dry, membranaceous expansion, flat extension, or appendage of an organ.
Woolly. Provided with long, soft, more or less matted hairs.

INDEX

[Names in italics indicate synonyms. Numbers in italics refer to illustrations.]

ABOY, 231
Acanthaceae, 14, 329
Acanthus family, 329
Achyranthes philoxeroides, 224
Acnida, 7, 13, 224
  cannabina, 224, 226
Acorus, 5, 6, 175
  calamus, 175, *176, 177*
  seedlings, *177*
Alisma, 6, 7, 78, 80
  geyeri, 80
  gramineum, 5, 79, 80
  plantago-aquatica, 4, 8, *79,* 80, *97*
  seedling, *97*
Alismaceae, 13, 78, 9~
Alligator weed, 224
Alopecurus, 7, 113, 114
  aequalis, 114, *115,* 116
  geniculatus, 114, *115,* 116
Alternanthera, 5, 7, 13, 224
  philoxeroides, 224, *225*
Amaranth family, 223
Amaranthaceae, 13, 223
American brooklime, 318
American lotus, 236
Anabaena, 350
Anacharis, 6, 11, 102, 104
  canadensis, 102, *103,* 104, 112
  densa, 2, 101, 102, *103,* 104
  occidentalis, 5, 8, 102, *103,* 104, *105*
  seedlings, *105*
Aquatic plants
  distribution, 3, 4, 5
  fruits, 6
  reproduction, 4
  seeds, 6, 7, 9
Araceae, 11, 12, 175
Arrow arum, *178*
Arrow grass, 76
  family, 72
Arrowhead, 86

Arrowroot family, 213
Arum family, 175
  water, 177
Awlwort, 253
Azolla, 349, 350, 351
  caroliniana, *347,* 351
  filiculoides, 351, *352*

BAKER, 352
Bald cypress, 2
Banana water lily, 244
Batrachium, 252
  aquatile, 249
  circinatum, 249
Beak rush, 161
BEATTIE, 255
Beckmannia eruciformis, 116
  syzigachne, *115,* 116
BEETLE, 2, 174
Beggar-ticks, 335
BENEDICT, 354
BENSON, 2, 252
Berula, 5, 7, 287
  erecta, 287
  pusilla, 287, *288, 293, 296*
Bidens, 7, 335, 338
  beckii, 4, *232,* 335, *336*
  bidentoides, 335, *336*
Black grass, 210
Bladderwort, 320
  family, 320
BLAKE, 189
Blue water lily, 244
Bluejoint grass, 116
Bog bean, 302
Bog moss, 191
  family, 189
BRACKETT, 174
Bramia, 6, 308
  monnieri, 308, *309, 310*

Brasenia, 6, 7, 231, 234
  schreberi, *233*, 234
Brass buttons, 338
Brook grass, 117
Brooklime, American, 318
BROOKS, 189
BROWN, C. A., 354
BROWN, W. F., 95
BUCHENAU, 95
Buck bean, 302
  family, 301
Buckwheat family, 213
BUELL, 181
Bulrush, 166, 167, 174
Bur reed, 18
  family, 18
Burheads, 82
Butomaceae, 13, 95
Butomus, 6, 97, 98
  seedling, 99
  umbellatus, 2, 8, 98, 99, 101
Buttercup, 247
  family, 246
  seaside, 248
BUTTERS, 72
Button rods, 192
Buttonbush, 2

Cabomba, 6, 13, 231, 234
  caroliniana, 5, *232*, *233*, 234
Calamagrostis, 7, 113, 116
  canadensis, *115*, 116
Calla, 5, 6, 175, 177
  palustris, 178, *179*
  wild, 178
Callitrichaceae, 14, 258
Callitriche, 7, 260
  autumnalis, 260
  bolanderi, 260
  hermaphroditica, 260, *261*, 262
  heterophylla, 260, *261*, 262
  palustris, 260, *261*, 262
  stagnalis, 260, *261*, 262
Caltrops, 274
Canary grass, 131
Carex, 7, 141, 174
  aquatilis, 141, 142, *145*
  comosa, 142, *145*
  inflata, 142, *144*, 146
  lanuginosa, 142, 143, *144*
  lasiocarpa, 142, 143, *144*
  leptalea, 142, 143, *145*
  limosa, 142, 143, *144*
  lyngbyei, 141, 143, *145*
  mertensii, 142, 143, *144*
  nudata, 141, *145*, 146
  obnupta, 141, *145*, 146
  rostrata, 146

  utriculosa, 146
  vesicaria, 142, *144*, 146
Caryophyllaceae, 14, 226
Castalia, 241
Catabrosa, 7, 114, 117
  aquatica, 117
Cattail, 15, 17
  family, 15
Celery, wild, 109
Cephalanthus occidentalis, 2
Ceratophyllaceae, 11, 227
Ceratophyllum, 6, 7, 228, 230, 231
  demersum, 4, 229, *230*, 231, 232
  echinatum, 229, *230*, 231
  seedlings, *230*
Ceratopteridaceae, 352, 354
Ceratopteris, 352
  deltoides, 352, 353
  pteridoides, 352, *353*
Chamaedaphne calyculata, 2
CHAMBLISS, 140
Chenopodiaceae, 13, 218
Chestnut, water, 160, 274, 276, 277
Cinquefoil, 258
Cladium, 141, 146, 147
  jamaicensis, 147, *163*, 164
  mariscoides, 147, *163*, 164
CLAUSEN, A. B., 78
CLAUSEN, R. T., 2, 27, 70, 72, 95, 352
Club rush, 166
Compositae, 13, 334
Composite family, 334
CONARD, 246
Coontail, 228
Cord grass, 134
CORE, 101
CORRELL, 354
COTTAM, 64
Cotton grass, 160
Cotula, 7, 335, 338
  coronopifolia, 3, *337*, 338
COULTER, 296
Crassulaceae, 13, 257
Cress, 253
  water, 255
CROCKER, 8
Crowfoot, 247
  water, 248, 249
Cruciferae, 14, 252
Cut-grass, 125
Cymodocea, 7, 27, 56
  manatorum, 57
Cyperaceae, 12, 140
Cyperus, 7, 141, 147
  articulatus, 147, 148, *149*
  ochraceus, 147, 148, *149*

Damasonium, 6, 7, 78, 82
  californicum, *81*, 82

DANDY, 56
DAVIS, 8
Decodon, 5, 6, 265
  verticillatus, 267, 268
Dianthera, 329
Dicotyledons, 13
Didiplis diandra, 269
DIEHL, 64
Distichlis, 5, 7, 114, 117
  spicata, 117, *118*
Distribution of aquatic plants, 3, 4, 5
Ditch grass, 59
Dondia, 223
DREW, 252
Dropwort, 295
Duckweed, 183, 185
  family, 181
Dulichium, 7, 141, 148
  arundinaceum, 148, *149*
DUVEL, 8

EAMES, 112
Echinochloa, 7, 113, 117, 119
  crus-pavonis, 119
  paludigena, 119
  walteri, *118*, 119
Echinodorus, 5, 7, 78, 82
  cordifolius, 82, *83*
  radicans, 82, *83*, 84
  tenellus, 82, *83*, 84
EDWARDS, 8
Eelgrass, 61, 64, 65
Eichhornia, 6, 12, 199, 203, 205
  crassipes, 2, *200*, 205, 206
  paniculata, 3, 205, 206
  seedlings, *200*
Elatinaceae, 14, 262
Elatine, 6, 262, 264, 265
  americana, *263*, *264*, 265
  californica, 264
  minima, 5, *263*, 264
  triandra, 264, 265
Eleocharis, 5, 6, 7, 141, 148, 150, 174
  acicularis, 151, *158*
  calva, 150, 151, *157*
  caribaea, 150, 151, 152, *155*
  cellulosa, 150, 151, 152, *154*
  diandra, 156
  elongata, 150, 151, 152, *154*
  engelmanni, 150, 152, *155*
  equisetoides, 150, 152, *154*
  intermedia, 151, 152, *158*
  microcarpa, 151, 153, *158*
  obtusa, 150, 153, *155*
  olivacea, 150, 153, *158*
  ovata, 150, 153, *155*, 156
  palustris, 150, 153, 156, *157,* 174
  quadrangulata, 150, *154,* 156
  reclinata, 152

  robbinsii, 150, *154,* 156
  rostellata, 151, 156, *158*
  smallii, 150, 156, *157*
  tuberosa, *159,* 160
  uniglumis, 150, 153, *157,* 160
Elm, water, 2
Elodea, 102, 104, 112
Elodea, see Anacharis
  planchonii, 104
Equisetaceae, 11, 345
Equisetum, 346
  fluviatile, 346, *347*
ERICKSON, 345
Eriocaulaceae, 12, 191
Eriocaulon, 6, 192
  compressum, 192, *193,* 194
  decangulare, 192, *193,* 194
  parkeri, 192, *193,* 194
  ravenelii, 192, 194
  seedlings, *195*
  septangulare, 192, *193,* 194, *195*
Eriophorum, 7, 141, 160
  virginicum, 160, 161, *162*
  viridi-carinatum, 160, 161, *162*
Eryngium, 7, 287, 288
  aquaticum, 288, 292
  prostratum, 288
Eryngo, 288
European brooklime, 318
Evening-primrose family, 269

False
  loosestrife, 272
  pimpernel, 316
Families, key to, 11
Fanwort, 234
FASSETT, 257, 265, 285
Featherfoil, 301
Fern
  floating, 352
  horn, 352
  water, 350, 351
Fernald, 2, 18, 21, 56, 61, 72, 95, 112, 140, 174, 223, 231, 265, 285, 296, 306, 332
Figwort family, 307
Fish grass, 234
Five-finger, 258
Floating
  fern, 352; family, 352
  heart, 302
Flowering quillwort, 72
Flowering rush, 98
  family, 95
Fluminea, 7, 114, 119
  festucacea, *118,* 119, 120
Foxtail, 114
  water, 114
Frogbit, 107
  family, 101

Galingale, 147
Gentianaceae, 301
Germination of seeds, 8
GILBERT, 189
Glasswort, 218
Glaux, 6, 298
　maritima, 298, 299
Glyceria, 7, 114, 120
　acutiflora, 120, 121, *123*
　borealis, 120, 121, *123*
　canadensis, 120, 121, 122, *123*
　elata, 121, 122, *123*
　fernaldii, 121, 122, *123*
　grandis, 121, 122, *123*
　leptostachya, 120, 121, 122
　melicaria, 120, 121, 122, *123*
　obtusa, 120, 122, *123*, 124
　occidentalis, 120, 122, *123*, 124
　pallida, 121, 122, *123*, 124
　pauciflora, 120, *123*, 124
　septentrionalis, 120, *123*, 124
　striata, 121, *123*, 124
GOEBEL, 189
Golden club, 178
Goosefoot family, 218
Gramineae, 12, 112
Grass family, 112
Grasses, 140
Gratiola, 6, 308, 311, 313
　aurea, 313
　lutea, *312*, 313, 314
　neglecta, *312*, 313, 314
　virginiana, *312*, 313, 314
　viscidula, *312*, 313, 314
GRAVES, 61
Great duckweed, 185
GREGOR, 332
GUPPY, 8

HAGSTRÖM, 56
Halodule, 7, 27, 57
　wrightii, 57, *58*
Halophila, 6, 102, 104, 106
　baillonis, 106
　engelmanni, 106, *110*
Haloragidaceae, 13, 14, 277
Hedge hyssop, 311
HEGELMAIER, 189
Hemianthus, 6, 308, 313
　micranthemoides, 314, *315*
Hemp, water, 224
Herpestis, 6, 308
　rotundifolia, 309, *310*
HERZOG, 352
Heteranthera, 6, 12, 199, 202
　dubia, *195*, *201*, 202
　limosa, *201*, 202
　mexicana, 202

reniformis, *201*, 202
seedlings, *195*
HICKS, 189
Hippuris, 7, 13, 277
　vulgaris, 5, 277, 278, 279
HITCHCOCK, 2, 140
Horn fern, 352
Horned pondweed, 61
Hornwort, 228
　family, 227
Horsetail, 346
　family, 345
Hottonia, 6, 14, 298, 301
　inflata, 298, *300*, 301
Howellia, 6, 332
　aquatica, 334
Hyacinth, water, 203
Hydrocharitaceae, 11, 13, 101
Hydrochloa, 7, 113, 125
　carolinensis, 124, 125, *126*
Hydrocleis, 6, 97, 98
　nymphoides, 2, 98, *100*
Hydrocotyle, 7, 287, 288, 289
　americana, 289, *290*
　bonarensis, 289, *291*
　ranunculoides, 289, *291*
　umbellata, 289, *290*, 294
　verticillata, 289, *290*, 294
Hydrolea, 6, 307
　affinis, *303*, 307
　quadrivalvis, *303*, 307
Hydrophyllaceae, 14, 306
Hydrotrida, 6, 308, 309
　caroliniana, 309, *310*
Hyssop
　hedge, 311
　water, 311

Isoëtaceae, 11, 339, 345
Isoëtes, 73, 339, 340, 345
　bolanderi, 340, 341, 342
　butleri, 340, 342
　eatoni, 340, 342
　echinospora, 340, *341*, 342, 345
　engelmanni, 340, *341*, 343
　flaccida, 340, 343
　flettii, 342
　gravesii, 342
　howellii, 340, 343
　macrospora, 340, 343
　melanopoda, 340, 343, 344
　melanospora, 340, 343, 344
　occidentalis, 340, 344
　orcuttii, 340, 344
　piperi, 344
　riparia, 340, 344, 345
　saccharata, 340, 344, 345
　tuckermani, 340, *341*, 345
IVERSEN, 345

Jaumea, 7, 335, 338
　carnosa, *337*, 338
JONES, 246
Juncaceae, 12, 206
Juncaginaceae, 12, 13, **72**
Juncus, 6, 73, 206, 207
　acuminatus, 207, *208*
　balticus, 207, *208*
　canadensis, 207, 210, *211*, 212
　gerardi, 207, *208,* 210, 212
　marginatus, 207, *209*, 210, 212
　militaris, 5, 207, *209*, 210, 212
　pelocarpus, 207, 210, *211*, 212
　polycephalus, 207, *209*, 210, 212
　repens, 207, 210, *211*, 212
　subtilis, 207, 210, 212
Jussiaea, 6, 269, 270
　californica, 270
　decurrens, 270
　diffusa, 270, *271*
　grandiflora, 270, *271*
　peruviana, 270, *271*
Justicia, 6, 329
　americana, 5, 329, *330*, 331

Key to families, 11
Knotweed, 213

Labyrinthula, 64, 65
Lake cress, 253, 255
LAKELA, 27
Leatherleaf, 2
Leersia, 7, 113, 125
　hexandra, 125
　oryzoides, 125, *126*, 127
Lemna, 7, 183, 185
　cyclostasa, 183, 184, *186*
　gibba, 183, 184, *186*
　minima, 183, 184
　minor, 4, 183, 184, *186*
　perpusilla, 183, 184
　seedling, *186*
　trisulca, 4, 183, 184, 185, *187*
　valdiviana, 184
Lemnaceae, 11, 181, *187,* 189
Lentibulariaceae, 11, 14, 320
Lettuce, water, 181
Lilaea, 7, 12, 72
　subulata, 73, *74*, 78
Lilaeaceae, 73
Lilaeopsis, 7, 287, 294
　carolinensis, 294
　chinensis, 294
　lineata, *293*, 294, 295
　occidentalis, *293*, 294, 295
Limnobium, 6, 102, 106
　spongia, 106, 107, *110*
Limnocharis, 6, 97, 101
　flava, **2**, 101
Limosella, 6, 308, 314, 316
　aquatica, *315*, 316
　subulata, *315*, 316
LINDBERG, 252
Lindernia, 6, 308, 316
　dubia, *315*, 316, 317
Littorella, 7, 13, 331, 332
　americana, 5, 331, 332
LLOYD, 329
Lobelia, 6, 332, 334
　dortmanna, 5, *330,* 334
　family, 332
　seedling, *330*
　water, 334
Lobeliaceae, 14, 332
Loosestrife, 268, 301
　false, 272
　family, 265
　swamp, 265, 268
Lophotocarpus, 7, 78, 84, 95
　californicus, 84, 86
　calycinus, *81*, 84, 86
Lotus, 9, 234
　American, 236
　yellow, 244
Ludvigia, 6, 269, 272
　alternifolia, 272, *273*
　linearis, 272, *273*, 274
　palustris, 272, *273*, 274
　sphaerocarpa, 272, *273*, 274
LUDWIG, 8
Luziola, 7, 113, 127
　bahiensis, *126*, 127
　peruviana, 127
Lysimachia, 6, 14, 298, 301
　terrestris, *299,* 301
Lythraceae, 14, 265
Lythrum, 6, 265, 268
　lineare, *266*, 268, 269
　salicaria, 3, *266*, 268

MACKENZIE, 2, 174
Macuillamia, 6, 308, 311
　repens, 311
　rotundifolia, *310*, 311
MAGUIRE, 257
MALME, 199
Manatee grass, 57
Mangrove, 2
Manisurus, 7, 113, 127
　altissima, 2, 128
Manna grass, 120
Marantaceae, 13, 213
Mare's-tail, 277
Marie-Victorin, 112
Mariscus, 146
Marsh cinquefoil, 258

Marsilea, 346, 348
  macropoda, 348, *349*
  quadrifolia, 3, 348, *349*
  uncinata, 348, *349, 350*
  vestita, 348, *349, 350*
Marsileaceae, 11, 346
MASON, 189
Matai, 160
Mayaca, 6, 189, 191
  aubleti, *190*, 191
  fluviatilis, *190*, 191
Mayacaceae, 12, 189
Menyanthaceae, 14, 301
Menyanthes, 6, 14, 302
  trifoliata, 301, 302, *303*
Mermaid weed, 285
MILLER, 246
Mimulus, 6, 308, 317
  guttatus, *312*, 317
Monanthochloë, 7, 114, 128
  littoralis, *126*, 128
Monkey flower, 317
Monocotyledons, 11
MOORE, 56
MORAN, 78
MORONG, 56, 78
MOUNCE, 64
Mud plantain, 199
Mudwort, 314
MUENSCHER, 9, 56, 77, 78, 101, 231, 255, 257, 277, 297, 332
Mustard family, 252
Myrica gale, 2
Myriophyllum, 6, 7, 277, 278, 285
  alterniflorum, 278, *282*, 283
  brasiliense, 3, 278, *281*, 283, 284
  exalbescens, 4, *232*, 278, *280*, 283
  farwellii, 278, *282*, 283
  heterophyllum, 278, *280*, 283
  humile, 278, *282*, 284
  proserpinacoides, 283
  scabratum, 278, *281*, 284
  tenellum, 5, 278, *280*, 284
  verticillatum, 278, *281*, 284

Naiad, 65
  family, 65
Najadaceae, 12, 56, 65
Najas, 6, 7, 65, 72
  conferta, 66, 70
  flexilis, 4, 66, *68, 70, 71*
  gracillima, 5, 65, 66, 67, *68*, 70
  guadalupensis, 66, 67, *68, 70*
  marina, 65, 67, *69, 70, 71*
  minor, 2, 65, 67, *69, 70*
  muenscheri, 66, 67, *69, 70*, 72
  olivacea, 5, 66, 67, *69, 70*, 72
  seedlings, *71*

seeds, 70
Nasturtium, 6, 252, 253
  officinale, 3, 8, 253, *254*
Naturalized aquatic plants, 2
Nelumbo, 5, 7, 231, 234
  lutea, 236, 246
  pentapetala, *235*, 236
Nuphar, 5, 6, 231, 236, 246
  advena, 237, *238*, 241, *245*
  americana, 241
  microphyllum, 237, *239*
  polysepalum, 237, *240*, 241
  rubrodiscum, 237, *239*, 241
  sagittifolium, 237, *240*, 241
  seedlings, *245*
  variegatum, 237, *238*, 241
Nymphaea, 6, 231, 241, 246
  elegans, 244
  flava, *243*, 244
  mexicana, 244
  odorata, 8, *243*, 244
  seedlings, *245*
  tetragona, *243*, 244, 246
  tuberosa, *242*, 244, *245*, 246
Nymphaea, 236, 246
Nymphaeaceae, 13, 14, 231
Nymphoides, 6, 14, 302
  aquaticum, 302, *305*, 306
  cordatum, 5, 302, *305*, 306
  peltatum, 3, 302, *304*, *305*, 306
  seedlings, *305*
Nymphozanthus, 236
Nyssa aquatica, 2

Oenanthe, 7, 287, 295
  sarmentosa, *292*, 295
OHGA, 246
OGDEN, 2, 56
Onagraceae, 14, 269
Ophiobolus, 64
Orontium, 6, 175, 178
  aquaticum, 7, *176*, *177*, 178, 181
  seedlings, *177*
Oxypolis, 7, 287, 295, 296
  canbyi, 296
  rigidior, 296

Panicum, 7, 113, 128, 129
  geminatum, 129, *130*
  paludivagum, 129, *130*
PARIJA, 200, 206
Parkeriaceae, 11, 352
Parrot's feather, 283
Parsley family, 287
Parsnip, water, 296
Paspalum, 7, 113, 129
  acuminatum, 129, 131
  dissectum, 129, *130*, 131
  repens, 129, *130*, 131

Peltandra, 6, 175, 178
　sagittaefolia, 178, 181
　virginica, 8, 178, *180,* 181, 205
PENNELL, 2, 320
Pennywort, water, 288
Peplis, 6, 14, 265, 268
　diandra, *267,* 269
Persicaria, 218
PFEIFFER, 2, 345
Phalaris, 7, 113, 131
　arundinacea, *130,* 131, 132
Pholiurus, 7, 113, 132
　incurvus, 2, 132
Phragmites, 5, 7, 114, 132
　communis, 132, *133,* 134, 140
　maximus, 140
Phyllospadix, 7, 27, 57, 59
　scouleri, 59, 62
　torreyi, 59
Pickerelweed, 203
　family, 199
Pilularia americana, 346
Pimpernel, false, 316
Pink family, 226
Pipes, 346
Pipewort, 192
　family, 191
Pistia, 6, 11, 175, 181
　stratiotes, 181, *182,* 183
Planera aquatica, 2
Plantaginaceae, 13, 14, 329
Plantago, 6, 14, 331, 332
　maritima, 332, *333*
Plantain, 331
　family, 329
　mud, 199
　seaside, 332
　water, 80
Pleuropogon, 7, 114, 132, 134
　californicus, 134
　refractus, *133,* 134
Podostemaceae, 13, 255
Podostemum, 6, 255, 257
　abrotanoides, 257
　ceratophyllum, *256,* 257
Polygonaceae, 13, 213
Polygonum, 6, 7, 213, 214, 218
　amphibium, 214, *216*
　coccineum, 214, 215, *216*
　densiflorum, 214, 215
　hydropiperoides, 214, 215, *217*
　portoricensis, 215
　punctatum, 214, 215, *217*
Pond cypress, 2
Pondweed, 27
　broad-leaved, 33
　family, 27
　horned, 61
　sago, 50

Pontederia, 7, 12, 199, 203
　cordata, 203, *204, 205*
　lanceolata, 203
　seedling, *204*
Pontederiaceae, 12, 199
Pool mosses, 191
PORSILD, 246
Potamogeton, 5, 6, 7, 27, 29, 56
　alpinus, 30, 32, 33, *35*
　americanus, 50
　amplifolius, 31, 33, *34, 52*
　angustifolius, 44
　berchtoldii, 8, *28,* 31, 33, *36,* 38, 54
　bicupulatus, 33, 38
　bupleuroides, 51
　capillaceus, 5, 32, *37, 38, 53*
　clystocarpus, 31, *38,* 39
　confervoides, 30, 38, 39, *40, 52*
　crispus, 2, 30, 38, *41*
　diversifolius, 5, 33, *37, 38,* 39
　epihydrus, 4, 32, 39, *42*
　fibrillosus, 31, 39
　filiformis, 5, 30, 39, 44, *47*
　foliosus, 30, 31, 39, *40,* 44
　friesii, 31, *40,* 44
　gemmiparis, 31, *37,* 44
　gramineus, 1, 4, 32, *43,* 44, 45, *53*
　hillii, 31, 44, 45
　illinoensis, 30, 32, 44, 45, *46*
　lateralis, 32, *37,* 45
　longiligulatus, 31, 45
　lucens, 44
　natans, 4, 32, *34,* 45, 50
　nodosus, 32, *46,* 50
　oakesianus, 32, 50
　obtusifolius, 31, *49,* 50
　panormitanus, 51, 54
　pectinatus, 4, 30, *47,* 50, 51
　perfoliatus, 30, *48,* 51
　porteri, 31, 51
　praelongus, 5, 30, *35,* 51
　pulcher, 32, *42,* 51, 54
　pusillus, 4, 31, *36,* 51, 54, *56*
　pusillus, 33, 54
　richardsonii, 4, 30, *48,* 54, 55
　robbinsii, 30, *41,* 54, 55
　seedlings, *28, 52, 53*
　spirillus, 5, 33, *43,* 54, 55
　strictifolius, 31, *36,* 55
　tenuifolius, 33
　vaginatus, 5, 30, *47,* 55
　vaseyi, 5, 32, *43,* 55, *56*
　zosteriformis, 4, 31, *49,* 56, *109*
Potamogetonaceae, 12, 13, 27
Potentilla, 7, 258
　palustris, 258, *259*
Primrose
　family, 298
　willow, 269

Primulaceae, 14, 298
Proserpinaca, 7, 277, 285
  palustris, 285, *286*
  pectinata, 285, *286*
Purslane, water, 268, 274

Quillwort, 339
  family, 339
  flowering, 72

Ranunculaceae, 13, 246
Ranunculus, 6, 7, 247, 252
  ambigens, 247, 248, *250*
  aquatilis, *232*, 249, *251*
  circinatus, 247, 249, *251*, 252
  cymbalaria, 247, 248, 249, 252
  delphinifolius, 248
  flabellaris, 247, 248, *250*
  flammula, 247, 248, 249, *250*
  hederaceus, 247, 248, *251*, 252
  lobbii, 247, 248, *251*, 252
  longirostris, 249
  purshii, 247, 249, *251*
  reptans, 248
  salsuginosus, 252
  trichophyllus, 249
Reed, 132
  canary grass, 131
  grass, 116
Reimarochloa, 7, 113, 134
  oligostachya, *133*, 134
RENN, 65
Reproduction of aquatic plants, 4
Rhizophora mangle, 2
Rice, wild, 138, 140
Riverweed, 255
  family, 255
Rorippa, 6, 252, 253
  aquatica, *232*, 253, *254*
Rosaceae, 13, 258
Rose family, 258
ROSE, J. N., 296
ROSENDAHL, 72
ROSSBACH, G. B., 329
ROSSBACH, R. P., 227
Ruppia, 7, 27, 59, 61
  maritima, 59, *60*, 61, *63*
  occidentalis, 59
  seedlings, *63*
Rush, 206
  bayonet, 210
  beak, 161
  club, 166
  family, 206
  flowering, 98
  spike, 148
  twig, 146
RYDBERG, 111, 112
Rynchospora, 7, 141, 161

  alba, 161, 164, *165*
  careyana, 161, 164, *165*
  cymosa, 161, 164, *165*
  macrostachya, 161, 164, *165*
  traceyi, 161, 164, *165*

SAEGER, 189
Sagittaria, 6, 7, 78, 84, 86, 95
  ambigua, 87, 89, 92
  brevirostra, 87, 89, *90*, 92
  cuneata, *85*, 87, 89, 92
  eatoni, 87, *88*, 89, 92
  engelmanniana, 87, 89, *91*, 92
  falcata, 87, 89
  graminea, 87, *88*, 89, 92
  kurziana, 95
  lancifolia, 87, 89, 92, *96*
  latifolia, 1, 4, 8, *85*, 86, *87*, 89, 93, *97*, 205
  longirostra, 87, *90*, 92, 93
  montevidensis, 2, 86, 93
  platyphylla, 86, 93, 95, *96*
  rigida, 87, *91*, 93, *94*, 95
  seedling, *97*
  subulata, 86, 93, 95, *96*
  subulata var. gracillima, 93
  subulata var. lorata, 93
  subulata var. natans, 93
  teres, 87, *88*, *94*, 95
  weatherbiana, 87, *94*, 95
ST. JOHN, 112
Salicornia, 7, 13, 218
  bigelovii, 218, 219, *220*
  europaea, 218, 219, *220*
  perennis, 218, 219, *221*
  rubra, 218, 219, *220*
  utahensis, 218, 219, *221*
Salix, 2
Salt grass, 117
Salvinia, 350, 351
  natans, 351
  rotundifolia, *347*, 351, *352*
Salviniaceae, 11, 350
Samphire, 218
Sand spurry, 227
Saw grass, 147
Scheuchzeria, 6, 72, 73
  palustris, 73, *75*, *76*
  seedlings, *75*
Scirpus, 5, 6, 7, 141, 166, 174
  acutus, 4, 166, 167, *172*
  americanus, 4, 166, 167, *168*
  californicus, 166, 167
  etuberculatus, 166, 167
  fluviatilis, 166, 167, *169*
  heterochaetus, 166, 170, *172*
  olneyi, 166, *168*, 170
  paludosus, 166, 170, *171*
  robustus, 166, 170, *171*
  smithii, 166, 170, *173*, 174

subterminalis, 166, 170, *173*, 174
torreyi, 166, *168*, 174
validus, 166, *172*, 174
Scrophulariaceae, 14, 307, 320
Sea blite, 223
Sea milkwort, 298
Seashore salt grass, 117
Seaside plantain, 332
Seaweed, 107
Sedge, 141
Seedbox, 272
Seedlings, see various genera
Seeds
　germination, 7
　number per lb., 9
　production, 6
　storage, 7, 8
　weight, 9
Semaphore grass, 132
SETCHELL, 63, 65
SHERFF, 338
Short-awn foxtail, 114
Sickle grass, 132
Sium, 7, 287, 296
　suave, 295, 296, 297
Slough grass, 114
SMALL, 95
Smartweed, 213
　water, 214
SMITH, 2, 95
Sparganiaceae, 12, 18
Sparganium, 5, 7, 18, 19
　acaule, 20
　americanum, 8, 18, 19, 20, 22, 26, 27
　androcladum, 18, 19, 20, 22
　angustifolium, 18, 19, 20, 23
　californicum, 20
　chlorocarpum, 19, 20, 21, *24*
　eurycarpum, 8, 18, 19, 20, 21, *25*, 26
　fluctuans, 18, 19, 20, 21, *23*, *109*
　glomeratum, 27
　greenei, 20
　inflorescences, 19
　minimum, 18, 19, 21, *25*
　multipedunculatum, 8, 19, 21, *24*, 26
　seedlings, 26
Spartina, 5, 7, 114, 134, 135
　alterniflora, 135, *136*
　cynosuroides, 135, *137*
　leiantha, 135, *137*
　patens, 135, *136*, 138
　pectinata, 135, *137*, 138
Spatterdock, 236
Spearwort, 248
Speedwell, 317
　water, 318
Spergularia, 6, 227
　canadensis, *226*, 227
Spike rush, 148

Spirodela, 7, 183, 185
　oligorhiza, 185, *187*
　polyrhiza, 4, 185, *187*
Spurry, sand, 227
STANDLEY, 246
STANFORD, 218
Starwort, water, 260
STEVENS, 64
Stonecrop family, 257
Suaeda, 7, 13, 218, 223
　linearis, *222*, 223
　maritima, *222*, 223
Subularia, 6, 252, 253
　aquatica, 5, *254*, 255
SVENSON, 2, 174, 262
Swamp loosestrife, 265, 268
Sweet flag, 175
Sweet gale, 2

Taxodium ascendens, 2
　distichum, 2
Thalassia, 5, 6, 102, 107, 112
　testudinum, 107, *109*, *111*
Thalia, 6, 213
　dealbata, 213
THOMPSON, 189
Tillaea, 6, 258
　aquatica, 258, *259*
Trapa, 7, 274
　natans, 3, 9, 160, *275*, *276*, 277
　seedlings, 276
Trapaceae, 13, 274
Triglochin, 7, 72, 76
　maritima, 76, 77, *78*
　palustris, 76, 77, *78*
　striata, 76, 77, *78*
Tule, 167
Tupelo gum, 2
Turtle grass, 107
Twig rush, 146
Typha, 5, 7, 15, 17
　angustifolia, *16*, 17
　latifolia, 4, *16*, 17
　truxillensis, *16*, 17
Typhaceae, 12, 15

Umbelliferae, 14, 287, 296
Utricularia, 6, 320, 321, 329
　americana, 329
　biflora, 322, *325*
　cleistogama, 326
　cornuta, 9, 321, 322, *324*, *328*
　fibrosa, 322, 323, *325*, *328*
　geminiscapa, 8, 321, 322, 323, *325*, *327*, *328*
　gibba, 321, 322, 323, *325*, *328*
　inflata, 321, 322, 323, *324*
　intermedia, 321, 323, *325*
　juncea, 322

Utricularia (*continued*)
  minor, 321, 323, *325*, 326
  olivacea, 321, 323, *324*, 326
  purpurea, 5, 8, 321, *324*, 326, *328*, 329
  radiata, 322
  resupinata, 5, 321, *324*, 326, 327, *328*
  seedlings, *327*
  seeds, *328*
  subulata, 321, *324*, 326, 327, *328*
  virgatula, 321, *324*, 326, *328*
  vulgaris, 4, 321, *325*, 326, *328*

Vallisneria, 6, 102, 107
  americana, 8, 107, *108*, *109*, 112
  neotropicalis, 109, 112
  seedlings, *108*
  spiralis, 112
Vegetative propagation, 4
Veronica, 6, 308, 317, 318
  americana, 317, 318, *319*
  anagallis-aquatica, 318, *319*
  beccabunga, 3, 318, *319*
Violet, water, 301

Wasting disease of Zostera, 64
Water
  arum, 177
  celery, 295
  chestnut, 160, 274, 276, 277
  clover, 346; family, 346
  cress, 253, 255
  crowfoot, white, 249; yellow, 248
  elm, 2
  fern, 350, 351; family, 350
  foxtail, 114
  hemp, 224
  hyacinth, 203
  hyssop, 311
  lettuce, 181
  lily, banana, 244; blue, 244; dwarf, 246; family, 231; sweet-scented white, 244; white, 241; yellow, 236
  lobelia, 334
  marigold, 335
  milfoil, 278; family, 277
  nut, 9, 274, 276; family, 274
  parsnip, 296
  pennywort, 288
  pepper, 215
  plantain, 80; crowfoot, 247; family, 78
  poppy, 100
  purslane, 268, 274
  shield, 234
  smartweed, 214
  speedwell, 318
  starwort, 260; family, 258
  violet, 301
  willow, 268, 329
Waterleaf family, 306
Waterweed, 102, 104
Waterwort, 262
  family, 262
WEATHERBY, 352
White water lily, 241, 246
WHITFORD, 56
Widgeon grass, 59
WIEGAND, 61, 112
Wild
  calla, 178
  celery, 109
  rice, 138, 140; Texas, 140
Willow, 2
WINNE, 275, 276, 277
Winter buds, 29
Wolffia, 7, 183, 185, 188
  columbiana, 185, *187*, 188
  papulifera, *187*, 188
  punctata, 185, *187*, 188
Wolffiella, 7, 183, 188
  floridana, *187*, 188, 189
  lingulata, *187*, 188, 189
  oblonga, 188, 189
Woody plants, 2
Wool grass, 160
Wrack, 64, 107
WYLIE, 112, 206

Xyridaceae, 12, 196, 199
Xyris, 6, 196
  ambigua, 196, 197, *198*
  caroliniana, 196, 197, *198*
  fimbriata, 196, 197, *198*
  iridifolia, 196, 197, 199
  montana, 196, 197, *198*, 199
  smalliana, 196, 197, *198*, 199

Yellow
  -eyed grass, 196
  lotus, 244
  water lily, 236
YOUNG, 65

Zannichellia, 7, 27, 61
  palustris, *60*, 61
Zizania, 7, 113, 138
  aquatica, 7, 8, 138, *139*, 140
  seedling, *139*
  texana, 138, 140
Zostera, 5, 7, 27, 61
  marina, 61, *62*, *63*, 64, 65, *109*
  seedlings, *63*
  wasting disease of, 64